SAP IN THE CLOUD:
MIGRATION
ROADMAP

insider
◄ BOOKS

insiderBOOKS is a line of interactive eBooks from the publisher of SAP-insider. With insiderBOOKS, you can evolve your learning and gain access to the very latest SAP educational content.

The team at insiderBOOKS has created a brand-new solution for the digital age of learning. The insiderBOOKS platform offers affordable (and, in many cases, free!) eBooks that are updated and released chapter by chapter to ensure that you're always reading the freshest content about the latest SAP technologies.

Each eBook provides more than what you can get from the typical textbook—including high-resolution images, cloud-based material (access your content anywhere!), videos, media, and much more. Plus, thanks to our generous sponsors, you get the opportunity to qualify for free eBooks.

insiderBOOKS is Reading Redefined.

Please visit our website for additional information:
www.insider-books.com

Andrew Hacket and Adam Bundy

SAP in the Cloud: Migration Roadmap

From the publisher of *SAPinsider*

SAP in the Cloud: Migration Roadmap
by Andrew Hacket and Adam Bundy

Published by Wellesley Information Services, LLC (WIS), 20 Carematrix
Drive, Dedham, MA, USA 02026.

Publisher	Melanie A. Obeid
Product Director	Jon Kent
Managing Editor	John Palmer
Acquisitions Editor	Jawahara Saidullah
Editor	Andrea Haynes
Copyeditor	Nicole D'Angelo
Art Director, Cover Designer	Jill Myers
Production Artist	Kelly Eary
Production Director	Randi Swartz

About the Authors

 **Andrew Hacket** is an IBM Cloud Engineer and Technical Business Development Executive at IBM Cloud. He has 25 years of experience in IT, including system integration, system hosting, and cloud computing where he has served as a senior program manager and IT architect for large, world-class enterprises. He can be reached at andrew.hacket@us.ibm.com.

 **Adam Bundy**, SAP Client Solution Executive at IBM, has more than 16 years of experience helping enterprise clients solve their toughest IT problems. Adam has spent much of his career working with SAP solutions in technical and functional areas. He can be contacted at ambundy@us.ibm.com.

About This Book

For those of you who have decided that SAP in the cloud makes sense for your organization, congratulations! This book is for you.

This book is a sequel to *SAP in the Cloud: An Executive Guide* published by insiderBOOKS. While that volume made the case for why you should move your SAP applications to the cloud, this book will explain how to do it.

We've written this book for a wide range of audiences. C-level executives will find a rich source of information about migrating to the cloud. Every chapter begins with a high-level overview before it dives into the nitty-gritty detail. Executives should feel free to skim through the chapters.

We've written this book for business owners of the SAP platform—business analysts, data analysts, and even marketing teams—to provide a business context that does not rely on technical jargon.

SAP implementation specialists, including integrators, project managers, and program managers, will also benefit from this book. We'll walk you through the steps of an SAP cloud implementation to provide the background of how to build a business, create a project team, and work with cloud providers.

Finally, we've written this book for SAP technical staff—admins, developers, and other SAP professionals—and will provide enough technical detail to enable you to move your SAP suite to the cloud with confidence.

Contents

Chapter 6   Phase Three: Build 89

Chapter 7   Phase Four: Transition............................ 107

CHAPTER 1

A Review of the Cloud

There is a good chance you've purchased this book because you have decided the cloud is where you need your SAP applications to be.

Perhaps your decision was driven by a need to innovate—to bring your operation into mobile and social platforms. Maybe you reached this decision as a cost savings move, to reduce IT capital expense and to shift your IT resources closer to where they should be, away from managing complex systems and back to their core competency of innovating your business. Or maybe your organization perceives that your current SAP landscape has become unsustainable in the long term and you need to reduce your exposure.

Whatever your organization's motivation, it is important that you identify a reason. Moving SAP systems to the cloud, while far less complex than the average SAP R/3 implementation back in the software's infancy, still requires adequate planning and resources. That's where this book comes in, providing a roadmap of the steps you need to take to get SAP on the cloud.

Like any other project, it is important to have a clear reason, executive buy-in, a coherent project team, and qualified partners who can make the transition easier.

Let's begin with a clear reason. Projects exist for three reasons: to reduce costs, to increase revenue, and to eliminate risk.

Before we take you through the process of justifying your project, let's spend some time getting to know SAP on the cloud. The better informed you are the stronger case you can make to move your SAP project forward.

Let's back up with a quick refresher as to what we mean when we say SAP "on the cloud."

What Is the Cloud?

The cloud is a model of computing in which a shared collection of servers, networks, databases, applications, and services can be provisioned. The following are three approaches to the cloud:

- The *public cloud* refers to virtual computing resources in an external Internet data center that are distributed across a large geographical area and are usually owned and operated by third parties, such as IBM or Amazon.

- The *private cloud* is similar, but is deployed within a customer's data center (or managed by a third party). It includes significant security features protecting the infrastructure from network threats and securing the SAP services delivery environment behind a firewall, and is usually managed by the provider's security experts.

- The *hybrid cloud* combines the public and private models. Hybrid models represent the fastest growing trend of new cloud-based SAP implementations. Some pieces of a solution may reside in a public cloud, such as software-as-a-service (SaaS) business applications that have a customer-facing component, or a platform-as-a-service (PaaS) that enables developers to create outward-facing applications.

Understanding IaaS, SaaS, and PaaS

Infrastructure-as-a-service (IaaS) is a cluster or clusters of physical servers of hardware and code, in this latter case, the hypervisor. The hypervisor manages the allocation of computing resources based on demand and enables virtualization. IaaS is the processing, storage, networks, and other computing resources that can be provisioned for operating systems and applications.

Software-as-a-service (SaaS) refers to network-based access to commercially available software in the form of services, such as Netflix and Gmail, or thin client product suites such as Microsoft Office 365 or Adobe Creative Cloud.

Platform-as-a-service (PaaS) is a blend of the two, a platform on which developers can build and deploy web applications on a hosted infrastructure. The operating system, run-time environments, source control repository, and all other middleware are contained in the technical solution stack.

Private, Public, and Hybrid Clouds

The majority of SAP customers today build on private clouds. The main perceived benefit of building on private clouds is control. Many SAP systems are mission-critical and it is not unusual for customers to be especially cautious. Private cloud customers want to take advantage of the cloud's efficiencies while keeping control of access and computing resources. Typical features of a private cloud include the following:

- Self-service interface to control services, to enable IT staff to quickly and efficiently provision, allocate, and deliver on-demand IT resources

- Highly automated management of resource pools, which provides flexibility for supporting computing capability to storage, analytics, and middleware

- Sophisticated security and governance designed for a company's specific requirements

Private clouds are not solely deployed within your company's data center. They can also be managed by a third party. The key element that separates them from hybrid or public clouds is that the network and servers upon which they operate are operated solely for your organization.

What private clouds lack in comparison to hybrid or public clouds is that they can usually be somewhat cumbersome. Your ability to quickly innovate and ramp up is limited by the talent of your IT team. Likewise, the size of your data and systems footprint is limited to the size of the infrastructure you have constructed and the related capital you have expended toward it.

A hybrid approach is the intersection of classic or traditional IT and the cloud. This approach is in its infancy, but is rapidly growing and will become the predominant approach to cloud computing for SAP.

One way to think about hybrid systems is that of systems of record (SOR) integrating with systems of engagement (SOE). For example, a company could run its customer relationship management (CRM) solution with all its customer records and market intelligence as an SOR, but have a customer-facing interface (SOE).

That dividing line between SORs and SOEs is the biggest conundrum in planning, sizing, and implementing an SAP-in-the-cloud project. Which enterprise activities and SAP processes will run in the cloud? Which resources should move to the cloud?

SAP HANA and the Cloud

We are in the midst of a revolution, if you will, of SAP cloud capabilities. SAP HANA, the in-memory system that has been revolutionizing SAP solutions, has now been joined by the SAP HANA Cloud Platform, an in-memory PaaS offering upon which SAP applications can be built, extended, and run.

This also is the platform that is ideal for SAP S/4HANA, the next generation of SAP suites and solutions. SAP is disposing of the traditional three-tiered model of database server, application server, and presentation server into a solid, integrated offering. The developers of SAP HANA have moved the business processes to the database layer and eliminated the aggregate database indexes.

With SAP HANA, rather than maintaining an individual database for every application along with point-to-point integration and long running queries, such as in batch mode, the new model as shown in **Figure 1.1** is to create a single database for the entire landscape, which requires no integration and real-time execution of queries.

From:
- One DB per application
- Point-to-point integration
- Long running queries, e.g. in batch mode

To:
- One DB per landscape
- No integration necessary
- Real-time execution

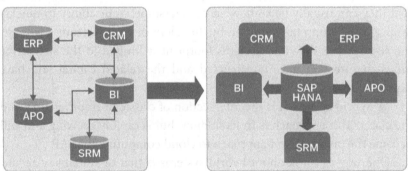

Figure 1.1 New SAP HANA model

SAP's cloud offerings come in three standard service classes: full service, development, and entry. Each conform to common SAP workloads in enterprise environments. Each service class includes SAP server infrastructure provisioning, operating system installation and licensing, database and SAP middleware installation, and configuration. Each service class varies in the degree of ongoing management and support services that are provided.

All the SAP Business Suite software applications can be run on any of the service classes. Each service class can also be provisioned in a range of different processing capacities, rated in SAP standard application benchmark units (SAPs), a standard measure of capacity for SAP systems. The difference in the service classes is in how they are intended to be used and the degree of ongoing support services that are bundled with each service class.

Service extensions are additional capabilities that can be added to the standard service class configurations when the system is provisioned. They can also be added to an existing, deployed SAP service to expand its capability. Service extensions also include special services that are typically requested to be performed for a single, limited duration on an existing SAP system or landscape.

Full Service Class

Full Service Class provides the full scope of SAP Basis Services required for the SAP-recommended best practice management for production SAP systems. Full Service includes provisioning, installation, configuration, and ongoing operations support for a production SAP system and two supporting non-production SAP systems for development and quality assurance testing. Full Service includes Start Up Services and SAP Basis administration services needed to support a production SAP system in a steady state, ongoing operations model. Ongoing services consist of SAP Monitoring, Basis administration for both ABAP and Java SAP updates, system and client management, output management, SAP online support services management, and automated SAP transport management. Database administration services, including periodic tablespace reorganization and troubleshooting, are part of the base service management. Monthly SAP Basis and database security patches are also part of the overall services needed to keep production systems stable and secured. SAP updates, database refreshes, SAP Basis functional patching, and manual transport management services are supported under a controlled and customer-initiated change management process under limited annual quantity entitlements.

Full Service SAP landscapes are available in multiple standard-sized rating capacities, ranging from 1,100 to 20,000 SAPs. Service extensions are supported to provide capacities spanning between the standard size capacities.

When an SAP landscape is mapped to a Full Service Class instance for provisioning, service extensions can also be applied to system sizing to increase the entitlements above the standard configuration for a number of parameters:

- Increase SAPs provided for individual SAP systems in the landscape

- Increase the amount of storage for databases in the landscape

- Increase the number of SAP systems in the landscape

- Add optional SAP components and auxiliary systems such as load balancers, portals, and SAP BusinessObjects servers

Development Service Class

Development Service Class is targeted to SAP workloads for a line-of-business user who wants to undertake a significant SAP project development activity and requires an environment not yet in production for developers and testers to run SAP functions. As with Full Service Class instances, when Development Service SAP systems are requested, a host provisions the SAP server infrastructure. A Development Service instance consists of an SAP landscape with a single non-production SAP system. Additional non-production or sandbox systems can be added as service extensions.

Development Service also includes installation and configuration of the SAP systems, onboarding, and ongoing management services for existing SAP systems. While the description of services for Full and Development Services are the same, the services provided have different focuses. Full service includes support processes to maintain local SAP systems in stable production. Development Service provides the more intensive support required by an SAP project development team.

Management services for Development Service Class instances are structured to support the needs and activities of the SAP development and test activities. As an example, Development Service is entitled to a greater number of database reorganizations and anticipates a greater frequency of SAP Basis functional patching. When the project hosted on a Development Service Class instance is ready to be put into production, an existing Development Service instance can be converted to a Full Service Class instance to support the transition to a production SAP environment.

Entry Service Class

Entry Service Class is targeted for SAP landscapes in which an SAP customer needs to run SAP systems for a short period of time and doesn't need maintenance or services for the SAP systems. Typical scenarios for Entry Service Class systems are for piloting new SAP applications, proof of concept modifications to existing systems, systems for training, or an exploratory application. Entry Service Class systems are not intended to be used for significant project development or testing extending over a longer period of time. Ongoing management and monitoring services are not provided for Entry Service because software components used to remotely monitor and manage Full and Development Service Class instances are not installed in Entry Service Class instances. For this reason, Entry Service Class instances cannot be converted into Development or Full Service Class instances.

Table 1.1 summarizes the services provided by IBM and SAP for each Service Class with indicators showing which services are provided for each Service Class.

Services	Description	Full	Dev	Entry
Start Up Services	**Architecture Services:** SAP-certified architect validates configuration before deployment	Yes	Yes	Yes
	Provisioning Services: Servers, storage, and network resources allocated to customer specifications from cloud infrastructure	Yes	Yes	Yes
	Infrastructure Configuration Services: Provisioned systems are configured with specific parameters and operating characteristics to support the specified SAP instances and landscapes	Yes	Yes	Yes
	SAP Database and Application Installation Services: Deployed according to IBM standard operating procedures and SAP installation guidelines	Yes	Yes	Yes
	Application Installation Services: Deployed according to standard operating procedures and SAP installation guidelines and documentation	Yes	Yes	Yes
	Infrastructure Configuration Services: Host configures operational infrastructure for the SAP landscape/instances	Yes	Yes	Yes
	SAP Configuration Services: SAP and database applications are configured according to SAP installation guidelines and host-developed best practices	Yes	Yes	Yes
	System and SAP Monitoring Setup: Monitoring agent is installed and configured into the IBM monitoring system	Yes	Yes	No
	System interface configuration and client configuration	Yes	Yes	No
	Automated SAP Transport Management: Sets up SAP transport management configuration, including workflow in accordance with Service Class specifications	Yes	Yes	No
SAP Basis Administration	Monitoring for SAP	Yes	Yes	No
	Maintenance of SAP profiles and configuration parameters	Yes	Yes	No
	SAP updates	Yes	Yes	No
	SAP system and client management	Yes	Yes	No
	SAP performance tuning (Basis level)	Yes	Yes	No
	SAP online support services (OSS) management	Yes	Yes	No
	Automated SAP transport management	Yes	No	No

Table 1.1 Service Class definitions

Services	Description	Full	Dev	Entry
Database Administration	Includes management of the SAP database(s), performing periodic reorganization, resolving database-related issues, and update management	Yes	Yes	No
Batch/Interface Monitoring	Provides job monitoring services for failed jobs and interface connections with an automated email notification process (for the Production environment)	Yes	No	No
SAP BusinessObjects Enterprise Administration	Includes administration activities supporting the SAP BusinessObjects applications and components; services include patching, instance management, output management, and performance tuning in support of current production functionality	Yes	Yes	No
SAP Service Reporting	This service provides monthly reports for service achievement against service level agreements and SAP system health review and performance/reliability improvements	Yes	Yes	No

Table 1.1 Service Class definitions

Application Size Classes

The processing and storage capacity required in an SAP system is often independent from the way the system is used. SAP sandbox or training systems derived from an existing production system can require sizable resources to support their intended usage. Conversely, a production SAP system can support a narrow business community with relatively small data requirements. For all the Cloud Managed Services service class options—Full, Development, and Entry—a second dimension of choice is provided for provisioning customer SAP systems: the Application Size Class.

The Application Size Class of an SAP service class instance determines the processing capacity for infrastructures provisioned for the SAP system. To provision systems with consistent performance characteristics across the choices of supported platforms, Application Size Classes are specified in terms of the SAPs and database storage allocation. Five application class sizes are available to be provisioned to support the resources required for an SAP end-user service instance. For Full Service Class instances, the SAPs ratings specify the throughput capacity provisioned for the production SAP

19

CHAPTER 1

system in the landscape. An equivalent number of SAPs are delivered for and divided between the development and quality assurance SAP systems also included in the landscape as shown in **Table 1.2**.

Entitlement Environment	SAPs		DB Storage (GB)		Memory (GB)	
	Prod	**Dev/QA**	**Prod**	**Dev/QA**	**Prod**	**Dev/QA**
XX-Small	1100	1100*	200	200/200*	7	12
X-Small	2200	2700*	400	300/400*	14	24
Small	3300	2700*	450	300/450*	20	24
Medium	6000	3000*	500	300/500*	36	24
Large	12000	6000*	600	300/600*	72	30
X-Large	20000	6000*	800	300/800*	120	30

Table 1.2 Application Size Classes

*SAPs for Development and Quality Assurance are shared across both systems.

In this book, we'll be giving you a personal view of the process of migrating your SAP systems to the cloud. As such, we'll explain the process of migrating SAP to the cloud from the standpoint of the business. While it would be helpful to have read our companion volume, *SAP on the Cloud: An Executive Guide*, it isn't necessary for much of the important information from that book has been summarized in this one.

As mentioned in the foreword, this book will have information that should be valuable to every level of an organization that touches SAP and is involved in a migration to the cloud. For CIOs who approach migrating to the cloud to solve challenges for the business, this book will offer some best practices for engaging stakeholders as well as making a business case for moving to the cloud. The CFO will be looking at the cloud to make sure that the financial resources devoted to IT are being put in the area that will generate the greatest return in the least amount of time at the lowest possible cost and without appreciably increasing risk—the heart of any project and there will be plenty of material that follows that will accomplish those goals.

The key benefit of moving to the cloud is that you will be moving to a lower cost, more agile platform. There are a lot of SAP systems out there and a lot of SAP customers have been looking to better control costs, complexity, and in many cases, contain growing technical environments. SAP on the cloud, specifically in a hosted environment, lifts that burden from SAP customers.

Another major benefit is flexible scalability. Your organization is changing faster than ever before. A decade ago, a merger or divestiture or acquisition created tremendous technical challenges for your IT organization. One of the great advantages of moving to the cloud is rapid provisioning and deprovisioning. Sharing, splitting, or growing your cloud-based SAP environment can now be done faster than ever and the cost is in the usage of the resource. No longer are customers bound by the size of their data center.

With only a few exceptions, most SAP customers now understand that the cloud environment and business model is mature.

Only a few years ago our clients would pepper us with questions with fears about security, access, ownership of data, and more. Eventually the major concern of executives became, "How much is this going to cost?" Today, we are heartened that we are receiving more and more queries related to how the cloud can promote business value in the form of faster development, faster implementations, rapid change of business processes and enabling new business processes.

From a technical standpoint, we also want to give readers a view under the hood. Many customers are quite advanced in their knowledge, and while this book is not written specifically for those with an active and thriving SAP presence on the cloud, you may find some information that will help you get more out of your current cloud-based operation, and certainly for any future implementation you may be considering.

For those who are beginning their cloud journey or considering moving SAP to the cloud, this book will serve an excellent introduction to the process and procedures behind such a move.

Over the course of this book you'll read about several topics where we talk about the value of SAP on the cloud, the proper steps to get on SAP HANA as well as the cloud. Also, we'll talk about the short- and long-term goals, the roadmap steps, you should consider to go through to get there.

So why the cloud and why SAP in the cloud? In many ways, it restores the original promise of enterprise architecture in that everything is about standardization. Standardization decreases complexity, allows for automation, optimizes resource allocation, and with the proper hardware and network resources in place, you enable scalability and resiliency.

A little more about resiliency: Because of the processes and data in an SAP system, or any other ERP system, resiliency has to be there. SAP is rapidly evolving, and it is becoming challenging to stay up to date on its developments. You need to have the specialists who know those changes and, more importantly, have the capability and experience to know the differences between ramp-up, prime-time, and mature. By moving to the cloud, those resources are now available to you rather than you having to expend your resources to keep your staff educated and up-to-date on the latest developments.

When it comes to migrating your SAP environment, in part or in whole, to the cloud, it requires a shift in mindset from a traditional IT approach. For decades, your typical SAP implementation required a significant investment, not only in hardware and software, but also in staffing needed to install, run, and maintain your SAP landscape. For SAP on the cloud, much of the emphasis on your own staff now shifts to your partner. Most SAP customers moving to the cloud will seek out a partner to host, integrate, and manage a company's core application environments.

Partners have always been important, either in the form of integrators or in SAP specialists. But by moving to the cloud, your partner becomes critical and is a significant key to achieve the business success desired.

IBM, as you know, is one of SAP's biggest partners on the cloud. We've hosted thousands of SAP customers and have done even more implementations. In our experience, the most successful migrations result from a joint client/IBM project team. The formula for success is simple: mix together your knowledge of your business with your cloud partner's extensive experience in migrating clients into hosted data centers. Together, this joint team balances consulting and operational resources to provide uninterrupted continuity of your business.

This service will allow you to transform your business application environment's workloads to leverage a virtualized, cloud-enabled infrastructure. Every cloud migration design begins with a thorough analysis of your environment and workloads. Based on the information we collect about your current environment, we can determine the numerous challenges you are dealing with and develop a cloud migration solution focusing on addressing these challenges. It leverages our experience with other clients in your industry, our experience with migrations, the skills and analytics we have used to study your challenges and IBM's point of view on the future of migration.

CHAPTER 2

Why Choose the Cloud?

For enterprises, moving to the cloud is no longer a question of if, but when. Three-quarters of you are already there in some form, taking advantage of the speed and scalability that the cloud offers. The popularity and benefits of the cloud do not mean that enterprises are abandoning traditional IT systems. The IT of the future, if we examine the trends, is that there will always be some form of hybrid IT landscape. Some recent reports predict that more than 80 percent of enterprise IT organizations will commit to hybrid cloud architectures by 2017.

Companies are choosing the cloud to deliver responsive IT and to innovate the way people work. Regardless of your role in your organization, you may find that cloud computing can open the door to business transformation. For corporate application managers, cloud is speeding the delivery of new products and services while also providing access to new services that can improve business processes. Managed services providers are using the cloud to open new revenue streams and offer differentiated services while reducing support costs. Chief information officers have been able to use the cloud to transform the responsiveness of their IT infrastructure and development efforts while improving management of compliance issues. For chief marketing officers, the cloud can provide a better place to store and analyze data more quickly, helping to reinvent customer relationships based on data, and expertise can be shared among customers, employees, and partners. Data center managers can use the cloud to provide self-service access to resources so that their staff can focus on higher-value projects, so important in a time when data center budgets strain the very best companies. **Figure 2.1** shows the benefits by role.

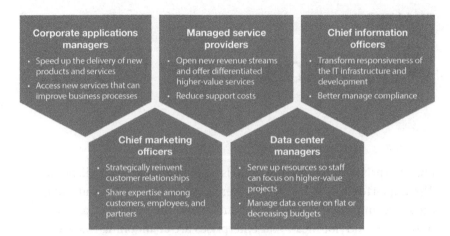

Figure 2.1 Benefits of SAP cloud computing by role

Many of the benefits of moving to the cloud are well known, among them the following:

- Reduce IT costs, such as space requirements, cooling systems, electrical and other power costs, and administration overhead.

- Encourage and enforce standardization of applications, infrastructure, services, and processes.

- Remove complexity, not only for users, but for your entire enterprise, especially in your SAP support areas, such as application managers and administrators dedicated to SAP Basis.

- Provide fine-grained services, tools, and options to enable sufficient customization and personalization without having to change or affect the core code.

- Aggregate data thanks to cloud-based analytical tools that can work equally well with structured and unstructured data.

- Shift capital expenses to operating expenses (or back, if that is your preference).

- Measure utilization to more precisely identify and cost IT services so that you only use the services you pay for, rather than building and maintaining capacity for the future.

While the cloud offers tremendous opportunities for growth and cost savings, it is not for the risk-averse. Cloud deployments introduce greater complexity in areas such as security and compliance, service and operations management, and cost management.

Many organizations are finding that cloud adoption has a greater impact on their cultural fabric, even more so than a business impact. The cloud is forcing companies to transform. For many early adopters there was a change management component that was not accounted for in the original planning. Many cloud projects failed because of the natural resistance to change that afflicts every organization.

The area most deeply affected is IT. We suggest that before you even begin your cloud project you take a hard look at your IT organization, especially to establish a baseline that later can be used to project how it will transform once you move to the cloud. You might want to ask yourself the following questions:

- How is IT supporting the enterprise?
- Is IT "at the table" shaping strategy?
- What IT initiatives are under way to support strategic initiatives?
- What are the line-of-business requirements and is your IT organization meeting them?
- Are lines of business pursuing "shadow IT" solutions, that is, IT solutions independent of corporate IT?

One of the end goals of any project of this magnitude is to generate an organizational harmony with the final result that IT, the business, and the cloud strategy work in concert rather than at cross purposes. Doing so will help you go beyond cost savings and flexibility to realize the full business value of the cloud.

This can be a tremendous opportunity for IT organizations that are struggling or out of sync with the business or with corporate strategy. Not only can it align IT with stakeholders, it can allow IT to be seen as a strategic adviser to business leaders to help them determine the best cloud services.

A 2013 survey by IBM of more than 800 executives found that even then getting on the cloud was a central objective. Many of those interviewed perceived the cloud as a mature, stable, and reliable solution for solving business challenges and increasing competitiveness. Interestingly,

CHAPTER 2

while cloud services were viewed as an area for cost savings, it also was believed to be a source for product innovation.

To build your business case for the cloud, you need to determine which role your organization will take in using and operating in the cloud. You can be one of the following:

- A cloud consumer, which is by far the most common answer
- A cloud provider, providing cloud services for others
- A cloud broker, the middleman between cloud consumers and cloud providers

For the purposes of this book we will only be examining projects for consumers of cloud-based services, but know that your cloud strategy may expand to include becoming a cloud provider or cloud broker.

What Are Your Business Drivers?

As we have emphasized, projects exist to support one or more of three axes: increase revenue, reduce costs, or mitigate risk. Migrating SAP systems to the cloud offers several potential benefits, including the following:

- Improves cost-efficiency by using economies of scale and helps shift capital expenditures to operational expenditures
- Reduces complexity, by using using automation and standardization
- Provides an enterprise-class infrastructure, and helps manage performance requirements
- Improves scalability, and provides an architecture designed to easily scale up or scale out
- Provides resiliency for core business processes; it uses a cloud-enabled infrastructure backed by clear key performance indicators (KPIs) and commensurate service level agreements

Based on those, common strategic and tactical objectives can be identified, including the following:

- Taking advantage of unstructured data from mobile sources or social media, and building analytics around it
- Streamlining test and development

- Transforming and expanding the capabilities of the data center
- Delivering innovative products and services more rapidly
- Gaining a robust foundation for transformational plays
- Running production workloads more cost effectively

Whatever your objectives, be they to reduce total cost of ownership, make IT more efficient, or deliver products to customers faster, what is important is to establish metrics and/or goals for each objective. Once you have established the business drivers and related metrics, then you have to determine what resources and infrastructure you need to achieve those objectives. This will require a gap analysis to take all the dimensions of your operating model and determine how far away you are from implementing the cloud properly to satisfy the business requirements attached to that major driver.

Defining Return on Investment (ROI)

The three project axes—reducing costs, increasing revenue, and mitigating risk—should always be front and center when estimating the ROI for any cloud project.

The common business case around moving to or starting a service in the cloud revolves around reducing IT budgets in one area—maintenance and operation, which can represent on average 70 percent of the IT expense—to increase spending on new initiatives where IT can contribute to revenue growth.

After thousands of cloud implementations, IBM has identified the four major areas where our clients have made and optimized ROI on their respective projects:

- **Standardization.** Simplify and streamline IT operations and minimizing customization.

- **Global or expanded delivery.** Expand into markets faster and more efficiently.

- **Information Technology Infrastructure Library (ITIL) processes.** Improve IT management processes to speed the cycle time and agility.

- **Economies of scale.** Scalability is at the heart of the cloud, where provisioning can happen almost with the push of a button.

Cost-Benefit Estimator

To help customers evaluate the value of cloud transformation projects, IBM offers the IBM Cloud for SAP Applications Benefits Estimator (see **Figure 2.2**).

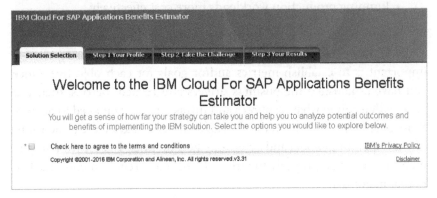

Figure 2.2 The IBM Cloud for SAP Applications Benefits Estimator enables you to calculate your ROI from migrating to the cloud

You can use the tool to outline your current IT environment and estimate what you would like to move to the cloud. The IBM Cloud for SAP Applications Benefits Estimator walks you through questions about your current IT landscape and how the cloud can improve it, using metrics such as speed to market, availability, and total cost of ownership.

Based on your input, and 15 minutes of your time, the tool will produce a report showing estimated savings from running SAP on the cloud. Check out the online tool at *http://ibm.co/2eZ273Q.*

The focus of this book is based on a "cloud-enabled" infrastructure. Most of you reading this book know that we are talking about SAP systems that were written long before there was a cloud, and its power is in the infrastructure, but now we have cloud-enabled them by adding self-service portals, pay-as-you-go models with rapid ramping up and down and scaling up and down.

Today, SAP cloud strategy centers around SAP HANA, which we discuss in detail in the Chapter 3. SAP HANA is not just a database, but a platform. When SAP launched R/3 in the early 1990s, it was supposed to be one system, with the comprehensive functionality required for your back-office systems. You migrated your data in there once and then you had one closed system that covered the entire enterprise.

Over the years SAP introduced new modules—SAP Business Intelligence (BI), SAP Customer Relationship Management (CRM), SAP Supplier Relationship Management (SRM), and SAP Advanced Planning and Optimization (APO), among others—and as you can see in the left half of the figure, these were connected but independent of SAP Enterprise Resource Planning (ERP). Now, with SAP HANA and the corresponding platform, SAP is pulling everything together again. All these applications can run on SAP HANA in separate instances today or together. In the future you will have third-party software providers write applications that also can take advantage of SAP HANA.

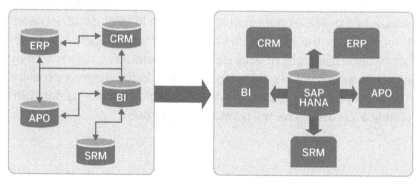

SAP from ERP to SAP HANA

In SAP HANA, the database is no longer separate that can be used for any kind of application or develop any kind of application on top. With SAP S/4HANA these systems will have similarly effortless integration giving you the best of both worlds, the comprehensive all-in-one of SAP R/3 with the ability to expand it into areas not even dreamed of today.

The SAP HANA platform was introduced in 2012, beginning with SAP Business Warehouse (BW). Because it was appliance based at the beginning, we were able to attract many customers to our cloud infrastructure at IBM. Now SAP HANA is running on more virtual machines and this has greatly increased cloud adoption. SAP HANA-powered applications run very well in a cloud-enabled environment for SAP.

One of the big selling points of the cloud is that SAP customers do not first have to migrate to SAP HANA and then migrate to the cloud. Your cloud partner simultaneously can convert you to the cloud and SAP HANA as one project. New customers to SAP are increasingly jumping in with SAP S/4HANA Finance, not only as an appropriate beginning point for SAP, but also because they have the confidence, thanks to SAP's long history in back-office systems, to know that the other modules will be coming.

What are the primary benefits? Not only can you reduce your IT expenses in this area (thereby enabling IT to focus on activities that grow the business), but also there is value in creating business processes that are not only faster but also in new areas that support the business.

Of course, cost is the primary driver. There is the cost of moving to the cloud, which we cover in a later chapter. But there is also the ongoing cost. This is determined by in what kind of cloud you want to operate. As the next figure shows, the trade-off between direct control and oversight of your cloud implementation versus turning it over to a cloud-managed service reflects on the ongoing expense. The latter is far less expensive. Also, the word "control" is not completely accurate, as we will point out in Chapter 7.

Cost

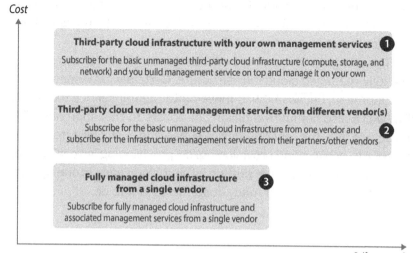

Self-managed

Comparing approaches for migrating SAP to the cloud

Migrating to the cloud can be done with a third-party cloud infra-structure partner managed by you. The advantages are multifold: You reduce your capital expenditures but at the same time have an infrastructure that can expand or shrink as needed by the business (such as seasonal fluctuations). One of the most important rea-sons cited is that you retain full control and flexibility in managing that infrastructure. The downside to this approach is that you must retain in-house skilled IT personnel as well as management of all the software licenses connected to the SAP environment. This approach also tends to discourage full automation in place of manual opera-tions to oversee the infrastructure.

Many SAP customers moving to the cloud prefer to work with a cloud infrastructure partner and another vendor to manage that cloud infrastructure. The benefit over the previous configuration is that it eliminates the need to have in-house IT infrastructure and management, thereby reducing capital and operational expenses. However, this approach creates a triage of responsibility, many times creating a "responsibility loop" where it can be difficult to determine

CHECKPOINT 2

who exactly is responsible for addressing an issue. Many SAP customers have had trouble finding an SAP consultant who has all the skills for the wide variety of SAP applications.

The area with the most growth has been SAP customers moving to a fully-managed cloud infrastructure. Certainly, SAP encourages this approach, because it has moved into the full-service model. At IBM we've also committed billions of dollars to this approach. The advantage over the other models is speed and cost. Full automation is easier to achieve and provisioning/de-provisioning is accelerated. The downside, of course, is putting all your cloud eggs in one basket, albeit this is exactly what many customers want to do.

CHAPTER 3

SAP on the Cloud: In Depth

SAP Cloud Platform is a unified and integrated platform-as-a-service (PaaS) solution that combines the traditional high performance, availability, and security metrics that SAP customers expect with the almost-real-time in-memory capabilities of SAP HANA.

The platform is flexible and is supported by a library of apps, with a growing number of services and features that can be deployed quickly and integrated with SAP or non-SAP infrastructures. The platform enables SAP customers to do the following:

- Extend and integrate SAP and non-SAP business processes into the cloud, including scenarios for mobile and the Internet of Things (IoT).

- Rapidly build and deploy cloud-based applications.

- Collect and analyze data from cloud apps and from on-premise SAP solutions in near real time.

- Create applications to engage customers, including taking orders and collecting revenue.

By running an SAP HANA system in the cloud (in this case on a hosted environment) vs. on-premise provides the benefits of no capital expense building data centers and no labor expense for dedicated staff to manage and maintain the system. There is no more managing patches and upgrades.

At the core of SAP's cloud approach is the enhanced performance and unique approach of SAP HANA, the in-memory core of SAP in the 21st century. Now there is SAP S/4HANA, which represents a complete rewrite of the software layer to take advantage of SAP HANA. To better understand and truly appreciate SAP S/4HANA, let's look under the hood of SAP HANA as a platform.

CHAPTER 3

The SAP HANA platform provides distinct services at the SAP HANA Application Services (SAP HANA AppServices), database (SAP HANA DBServices), and infrastructure (SAP HANA Infrastructure Services) levels.

SAP HANA AppServices enables the creation of consumer-grade, customer-facing applications that enable real-time analysis with SAP-level security. With SAP HANA DBServices (see next section), it is available in standard configurations ranging from 64 gigabytes to 12 terabytes. It includes shared services for application management, systems management, administration, and monitoring, as well as supporting a variety of analytics, mobile, portals, and collaboration features. SAP HANA AppServices is SAP-certified, and certified in general to meet industry standards for building, deploying, and managing mission-critical, cloud-based enterprise business applications. It also has native connectivity to back-end enterprise business content, including content from SAP and third-party solutions, whether running in the cloud or on premise.

Why SAP HANA?

SAP HANA combines speed and agility to support your most important workloads with faster transaction speed and response to new customer demands. SAP customers report that they operate at a higher level than ever before, with the ability to deliver new levels of service, integrate data more efficiently and faster, and gain actionable insights with a solution that fits their unique needs.

Other benefits include increasing the speed of business processes, simplifying IT transactions, and predicting and analyzing data simultaneously. From a business and ROI standpoint, customers are able to run multiple SAP HANA instances on one system and improve security, resource sharing, scalability, and workload isolation. This enables the delivery of new levels of service, integration with partners, and achieving actionable insight with a solution that fits anyone's unique needs.

SAP HANA DBServices comes in subscription-based configurations from 128 gigabytes to the same maximum as SAP HANA AppServices of 12 terabytes. It delivers provisioning of SAP HANA and hardware, and allows customers to build real-time analytics applications using the development capabilities of SAP HANA. For configuration and administration there is an intuitive cloud management console interface.

CHAPTER 3

SAP HANA DBServices is appropriate for a wide variety of use cases along a variation of axes including analytical, transactional, innovative Big Data, text, predictive, and data exploration. SAP HANA DBServices provides a secure environment in which each customer is assigned a virtual local area network (VLAN) and all services are covered under the same account and the same-user ID/password combination. All stored data is protected by SAP HANA security protocols (authentication and authorization), enabling customer-driven encryption for data transfers if that is desired.

SAP HANA Infrastructure Services enables customers to deploy and manage SAP HANA instances on the cloud and is available in configurations ranging from 128 gigabytes to 12 terabytes. Also included is the SAP Data Services component of SAP HANA Cloud Integration, which provides integration with SAP on-premise and third-party applications.

SAP HANA Infrastructure Services supports the provisioning of the SAP HANA platform, as well as SAP HANA Cloud Integration for data services, a cloud-based tool to extract, transform, and load data from SAP Business Suite and other sources, such as third-party databases, into SAP HANA. SAP HANA Infrastructure Services are appropriate for the following:

- Analytical and transactional use cases

- Innovative big data, text, and predictive use cases

- Data exploration use cases

Customers who require more than 12 terabytes of memory need to check with their integration and/or hosting partner. IBM, for example, supports larger data volumes. Options include the following:

- Build your own data center to host your cloud.

- Run SAP HANA on your desired platform (P or X).

- Using SAP's cloud, called SAP HANA Enterprise Cloud. SAP, as well as partners such as IBM, operate numerous data centers in the U.S.

- Use one of the SAP HANA cloud providers certified by SAP (see sidebar).

CHAPTER 3

SAP Certification for Cloud Providers

Cloud providers must be SAP certified. When it comes to certification for cloud providers, here are some of the raw criteria:

- You must be global, not only because of networking, transmission, and latency, but also because many countries, notably Germany and Switzerland, require that any data generated in that country must remain in that country.

- You have to be able to serve as a platform for other applications. Many SAP customers are also customers of other enterprise software vendors.

- You must have SAP HANA skills, expertise, and experience.

On-Premise or Cloud

Cloud-managed SAP solutions provide assets as services (data center, hardware, storage, monitoring software, and automation), labor (infrastructure, database, and SAP applications specialists), and processes required to manage and run a comprehensive and scalable SAP solution.

By avoiding up-front capital investment and ongoing maintenance costs typically associated with implementing a responsive SAP solution, the onus falls upon the partner to provide an enterprise-class, high-security cloud platform.

The hosting partner takes full responsibility for managing the technology and processes that deliver the application during activation and production and manages and controls production across the various layers of the solution. Best of all, from an expense standpoint, the SAP customer is charged based on usage, paying for the level of service that the business requires, without any capital investment in hardware or infrastructure. Among the various services cloud partners offer are the following:

- SAP infrastructure deployment and installation
- Operating system and server management
- SAP Basis configuration, tuning, administration, and support
- Storage and backup management
- Database management
- Data center and network management

Table 3.1 aligns the issues faced in on-premise, dedicated hosting environments with the operational advantages using cloud-managed services.

On-premise hosting	Cloud-managed services
New SAP systems' infrastructure requires significant capital investment	New SAP systems are provisioned from a shared host cloud services delivery model in which SAP services are virtual, billed monthly, and delivered as a service, reducing the need for capital
SAP-qualified staff consumed with routine SAP maintenance and operations, which limits their ability to support new critical SAP projects required for business growth	Automation of common SAP maintenance and operations tasks reduces SAP Basis and DBA administration workload complexity, freeing up resources for business-critical SAP development support
SAP development systems sit idle between development phases, continuing to incur capital payments, licensing charges, and maintenance	Unused development and test systems can be saved and de-provisioned from the cloud, eliminating operational costs and lowering the overall total cost of ownership for SAP workload environments
Custom infrastructure and SAP configurations require specialized staff training and administration skills to maintain unique configurations	Standard SAP images are built to predefined architecture templates and provisioning and configuration services are largely automated due to standardization. This approach reduces the need for costly staff training for custom configurations while also improving provisioning speed and quality
Underutilized non-production SAP systems consume customer resources maintaining SAP Basis and databases	Client resources and funding utilized to maintain multiple SAP systems can be applied to only those systems actively being used, providing better utilization of both staff and capital resources
Customer requires global support/skills and scalability capabilities to accommodate growth, or the customer has limited SAP skills needed for expanded SAP services demands	Many hosting partners manage global SAP systems in complex demanding environments. The onus is on the partner to maintain a stable of highly skilled SAP practitioners. SAP services are delivered as a service and address the routine and intractable SAP migration, deployment, and operational tasks associated with expanding service demands
Improve quality and service-level commitments for production SAP systems. Reduce errors associated with provisioning and updating the SAP system, applications, and infrastructures.	SAP systems are deployed from standardized, validated system images tested against patch levels and a standardized infrastructure. The automation of deployment and configuration procedures eliminates most sources of error leading to quality and performance issues. Services are also backed up by SLAs for performance and availability

Table 3.1 Comparison of on-premise hosting and cloud-managed services

CHAPTER 3

On-premise hosting	Cloud-managed services
Difficulty managing complex SAP landscapes	Complexity in SAP landscapes arises from the evolution of an IT infrastructure acquired from different sources and customized for specific needs. A cloud management partner employs a reusable, shared infrastructure and standardized SAP components. Standardization provides the basis for simplifying and automating operations and life cycle processes complicated by the existing custom infrastructure
SAP customers need increased agility responding to changes in the business environment, but find long lead times to implement new SAP-based business process innovations	Rapid provisioning of SAP system clones in reusable cloud infrastructure enables clients to increase speed to market for new SAP implementations. The ability to rapidly clone existing SAP systems enables the use of agile development methodologies to modify existing systems and rapidly prototype changes to business processes
Access to proper development systems is limited, constraining the ability of development and test organizations to increase speed to market	With development and testing, SAP systems are assigned from a pool of computing resources shared across the enterprise. Development and test organizations will no longer need to secure dedicated resources to provide the availability of hardware for their goals. Development systems can be released back into the general pool at the conclusion of the development and test periods, rather than sitting idle until needed for the next development cycle, resulting in better resource utilization
Unreliable incident, problem, and change management processes	Incident, problem, and change management is standardized and automated across the SAP services to improve repeatability, reliability, and quality, while also improving process speed
Following upgrades and technology refreshes, SAP systems provide unstable and unreliable service	SAP standard images and automated provisioning of SAP templates into the cloud eliminate common sources of error in SAP deployments and upgrades. Pretested standard templates provide deployed systems that have been thoroughly tested before they are deployed for production
Remove IT complexity from application managers	Standardized SAP images built on a virtual cloud infrastructure remove visibility to complexity into the infrastructure. SAP is implemented on a standardized system architecture image with management of the infrastructure provided by the host
Reduce errors and misconfigurations associated with SAP technology and database refreshes and updates	SAP standard images and automated provisioning of SAP images eliminate common sources of error in SAP deployments

Table 3.1 Comparison of on-premise hosting and cloud-managed services

On-premise hosting	Cloud-managed services
Technology refreshes and upgrades are complex and labor-intensive	Services are delivered in a shared platform as a service delivery model. The host owns and manages the SAP infrastructure technology in a shared delivery model and manages the client's dedicated SAP application environment on the shared infrastructure on behalf of the client, eliminating the need for the client to perform infrastructure technology upgrades/refreshes and standardization

Table 3.1 Comparison of on-premise hosting and cloud-managed services

SAP HANA on-premise versus in the cloud begins with the optimized server preloaded with SAP software components. The SAP HANA appliance is certified and comes in sizes that allow you to scale up by adding memory or scale out by adding services.

In the partner environment, a cloud provider can offer the full-blown, preconfigured SAP HANA solution or split off some key pieces of SAP HANA. In the latter case, the SAP HANA tailored data center integration (TDI) has the storage disengaged from the appliance. Obviously, because we are talking about SAP HANA, there is still an in-memory database system, but partners can provide the network and the storage (see **Figure 3.1**).

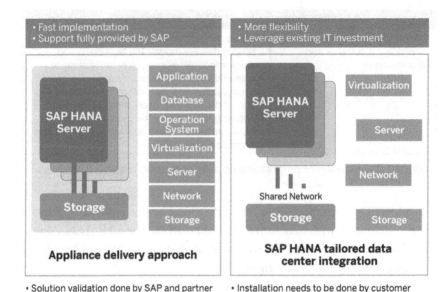

Figure 3.1 An appliance-delivery approach vs. SAP HANA tailored data center integration

With the SAP HANA tailored data center integration approach, there are restrictions and requirements, including the following:

- The installation must be SAP-certified
- SAP requires that the installation be tested with the SAP HANA hardware configuration check tool, a script that tests the system and then sends the test results to SAP for verification

Sizing SAP HANA

SAP prices SAP HANA solutions based on "T-shirt" sizes. This occasionally has resulted in confusion by SAP customers who are familiar with other pricing options related to web hosting. SAP HANA is not priced on consumption, as with other cloud services, but on what size SAP HANA T-shirt you require (see **Figure 3.2**).

Note: This is for scaling up. For scaling out, such as for SAP BW reporting systems, the SAP HANA nodes can only go up to 2TB.

T-Shirt Sizes Available for All Consumption Models

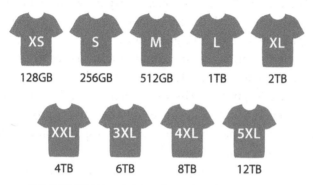

Figure 3.2 SAP HANA pricing is based on T-shirt sizes

Sizing Your System

SAP has a number of mature sizing tools for estimating how large your SAP HANA system will need to be.

NOTE: A ballpark ratio of the sizing of a database into SAP HANA is 3:1 vs. 2:1 for SAP S/4HANA. Note that this depends on your current level of data compression.

The general rule when sizing an SAP HANA system on-premise is that your system should be built to be 100 percent larger than what your sizing estimate indicates. Therefore, you are building based on a 50 to 60 percent utilization. The reasoning behind such conservative calculations is that the expense of adding to the physical infrastructure in the future can be considerable.

One of the benefits of locating SAP HANA in the cloud is that there is far more tolerance, or at least lower risk, for errors in sizing. An error may result in an increase per month in expanded usage costs versus hundreds of thousands of dollars in capital costs to expand an on-premise solution.

CHAPTER 3

Mission-Critical on the Cloud?

Moving mission-critical systems to the cloud for many SAP customers can be nerve-wracking. SAP HANA production databases of 12 terabytes and smaller can be migrated with relative ease (and comfort of mind) to the cloud, but larger systems should rightfully give one pause. The real risk, though, is in the changing business world we live in. Choosing to build your own data center, for example, can be expensive from a capital standpoint. What happens if the business changes? A divestiture, a sale, a merger—those capital costs will not be going away.

Larger SAP implementations on the cloud in a hosted environment are a hedge for that. But part of your due diligence in determining if the cloud is the right place to put SAP HANA is determining if your cloud partner can back up claims in its SLAs covering availability (production and non-production), performance, service delivery response and resolution times, disaster recovery, and high availability.

The biggest concern about moving mission-critical systems to the cloud is security. SAP Cloud Platform contains SAP's own search over encrypted data (a framework known as SEEED), which encrypts sensitive data.

SAP's cloud portfolio, including that of its certified partners, promises robust security controls at every level, including the following:

- **Data security and privacy.** SAP adheres to the European General Data Protection Regulation across all its software and systems, regardless of where they are located. If a country has stricter laws or regulations regarding data security and privacy, SAP meets or exceeds those requirements in that locale.

- **Security and compliance enforcement.** SAP products contain telemetry that tracks security and compliance from the time a product is developed until it is in production. That same telemetry warns customers when requirements are deviated from, as well as automating triggers to respond.

- **Physical security at SAP data centers.** Every SAP data center, as well as those of its partners, is located in secure, environmentally-controlled facilities with integrated security management. All SAP data centers comply with the latest telecommunications industry standards, such as American National Standards Institute (ANSI)/Telecommunications Industry Association (TIA)/Electronic Industries Alliance (EIA), 942 Tier III or higher.

- **Data storage and location.** In the cloud, data sources can be diverse. Segregating data, as well as continually sourcing it, becomes paramount, especially in light of privacy and security laws. SAP software contains controls that meet regulatory requirements for identifying data. SAP cloud products support logical isolation of data within a solution that extends to the virtual server layer.

SAP has a reputation for controlling user access to data and other information based on well-defined rules that are configurable based on a customer's security requirements. The cloud is no different. SAP's "least privilege" approach extends into its cloud portfolio to limit any given user's access to the minimum required to perform a set function. These security measures apply across the board and across all layers and assets, and enable SAP to meet strict regulatory requirements, such as International Organization for Standardization (ISO) 27001, International Standard on Assurance Engagements (ISAE) 3402, and Statements on Standards for Attestation Engagements (SSAE) 16.

At the same time, SAP's cloud portfolio contains tools to prevent, identify, and track any potential exploitation of technical vulnerabilities. These protocols are constantly reviewed by SAP and its partners with penetration tests conducted many times over the course of a year.

Security remains one of the biggest areas of focus by our clients and the questions they ask regarding the cloud environment include the following:

- Does it provide role-based security?

- Does it support secure, identity-based access from any location or device?

- Does it support user-activity reporting for auditing (including inactive accounts)?

- Does it provide the ability to deny service down to the IP address level?

- Can it integrate or support virtual firewalls?

- Will it integrate with governance, risk, and compliance platforms?

The answer to all these questions, as in any SAP implementation, should be yes. Later, we will address the elephant in the room, which is not the security properties of an SAP system, but how your hosting partner

CHAPTER 3

or your internal staff will manage security. Tools for security are only as good as the protocols, policies, and procedures of the staff that uses them.

Methodology of Moving

The methodology of moving to the cloud varies depending upon exactly what you want moved. Are you already an SAP customer? Do you already have SAP HANA? Is this a fresh SAP implementation?

Those moving an on-premise SAP HANA deployment to the cloud need to employ a tried-and-true methodology that works specifically for the on-premise-to-cloud process. See **Figure 3.3** for an example of the methodology we use.

Deployment phases	Launch	Model	Build	Transition
Key events	▪ Establish project team ▪ Establish project governance ▪ Develop project plans ▪ Review scope and transition ▪ Define priorities ▪ Define roles and responsibilities	▪ Infrastructure requirements and design—network, storage, backups, monitoring ▪ Application and business process discovery—architecture, interfaces ▪ Operational run books ▪ Migration strategy	▪ Provision shared infrastructure ▪ Build and test infrastructure ▪ Build and migrate databases, applications ▪ Phase cutover—development, quality assurance (QA), stage production	▪ Post-live technical support ▪ System stabilization and monitoring ▪ Operations process training ▪ Transition to IBM operations

Figure 3.3 IBM's standard process to support the transition and transformation of SAP environments and operations to the cloud

If you currently have an SAP implementation but are upgrading to SAP HANA, the process is more complicated (see **Figure 3.4**). For example, if you are going to move your on-premise SAP BW solution to SAP BW powered by SAP HANA in the cloud, it's a two-step migration. You have to perform the SAP BW upgrade followed by the SAP HANA upgrade. SAP has an automated tool, the Database Migration Option (DMO) of the Software Update Manager (SUM), that makes that two-step process easy.

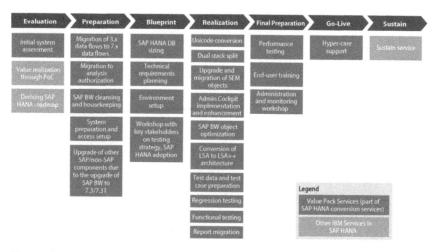

Figure 3.4 An effective migration plan allows customers to exploit new SAP HANA capabilities

Migrating SAP S/4HANA

Migrating SAP S/4HANA is trickier and only a handful of integrators and cloud providers have considerable experience. Before making the commitment to moving your SAP S/4HANA system to the cloud, you must make a special effort to check the following:

- Complexity of the upgrade
  - Source system version
  - OS and DB upgrade requirement
  - Level of customization
  - Unicode migration requirement
  - Regression testing requirement
- Size of the database
  - Migration to SAP Business Suite powered by SAP HANA
  - Migration to SAP S/4HANA Finance data structure
  - Database-specific features should be turned on
- Business downtime window
  - Distribution vs. industrial
  - Number of connected apps and interfaces
  - Number of trial cutovers to optimize the downtime

CHAPTER 3

What makes a good candidate for moving to the cloud? You're a good fit when you want to shift your IT infrastructure costs from a capital expense to an operational expense. You're also a good fit if you are moving to a virtualized landscape or are virtualized already. If you're looking for standardization vs. customization, the cloud is ideal for you. If you are looking to reduce operational risk, you shift that risk onto your cloud partner, manage it via service-level agreements (SLAs), and deliver managed service offerings for SAP solutions through a cloud model.

Your organization is an excellent candidate for the cloud if this is where you want to make the jump to SAP HANA or to SAP S/4HANA. And as mentioned earlier, if you are a company heading toward a merger/acquisition/divestiture, or if changing your portfolio of companies is a frequent occurrence, the move to the cloud reduces risk and complexity during those transition periods.

Moving SAP to the cloud is not for everybody. If your financial model is oriented more to capital expenditures for IT, the cloud may not be a good fit. If you are on an older version of SAP (hello to all our R/3 fans out there!) and have no interest in upgrading to a version that SAP currently supports, then the cloud is probably not the right fit. If you have large, standard costs in your existing infrastructure that you have to get off your books before such a move, the cloud may not make financial sense at this time.

None of these inhibitors, by the way, rules out the cloud, but before making any kind of leap you should consult with cloud experts who can help you calculate the benefit of such a move in light of these constraints. We see many more customers that can fit on the cloud than those that can't.

Before you even start an SAP on the cloud project, do your due diligence and engage a cloud partner to help research your options.

It is important that the cloud migration be part of your company's overall SAP application roadmap and that the roadmap take into consideration other transitional program-related activities on the application level. In many cases the SAP HANA migration is a technical project to enable the SAP systems users to leverage SAP

49

HANA, and the transformational changes in reporting and business processes such as SAP S/4HANA Finance are implemented in a later phase of a transformational program. That said, the anticipated end state of the applications should be considered when planning the SAP HANA migration so that all applications that require SAP HANA are migrated to the new database platform.

During the migration planning it is important to have an integrated plan with your provider as well as with the application support organization so that any required integration-related remediation activities for customizations are part of the overall plan.

An SAP HANA migration often requires the SAP applications to be upgraded to a more recent level, which creates an opportunity to get the SAP application suite modernized and ready for additional business process improvements.

Once the application plans and the timing for the SAP HANA migration are set, one needs to follow a methodology to model and build the cloud-based infrastructure capacity and then transition the SAP systems to the cloud environment.

In later checkpoints, we'll be using an example from a recent customer to explain and illustrate the steps covered in the following four chapters. The example we will use is an existing SAP customer (we won't mention names) sitting in a managed cloud environment and planning an ERP Central Component (ECC) and SAP BW database migration to SAP HANA and an implementation SAP S/4HANA Finance. In this example, the SAP HANA database migration itself is non-transformational and is meant to provide the technology platform to transform reporting and finance processes as a second phase.

CHAPTER 4

The Launch Phase

At the heart of any project, as we've stated earlier, is the intention that it will result in an increase in revenue, a reduction of expense, or mitigation of risk. However, there are challenges to be faced, as shown in **Figure 4.1**.

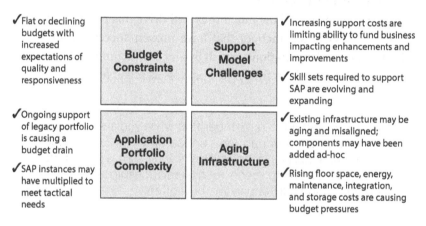

✓ Flat or declining budgets with increased expectations of quality and responsiveness

Budget Constraints

Support Model Challenges

✓ Increasing support costs are limiting ability to fund business impacting enhancements and improvements

✓ Skill sets required to support SAP are evolving and expanding

✓ Ongoing support of legacy portfolio is causing a budget drain

✓ SAP instances may have multiplied to meet tactical needs

Application Portfolio Complexity

Aging Infrastructure

✓ Existing infrastructure may be aging and misaligned; components may have been added ad-hoc

✓ Rising floor space, energy, maintenance, integration, and storage costs are causing budget pressures

Figure 4.1 The shift to the cloud often is driven by one of these four challenge areas

Projects, using a proper methodology, are how we address implementing innovations to achieve those objectives of increasing revenue, reducing cost, and eliminating risk.

Let's add a few words about project management. What we will not be doing in this book is giving you a primer on project management. The assumption is that you have your own methodology and personnel for that. What we will be focusing on is helpful advice and best practices for SAP migrations to the cloud.

CHAPTER 4

Onboard to the Cloud, Part One

Onboarding is the process of incorporating a cloud management partner into your project. This relationship is by far the most critical to the
success of any SAP cloud endeavor. While different partners may handle
onboarding differently, the best way to look at it is that your partner's
activities should mirror the staffing and steps that you will take with your
own project.

As with any project, you will assemble a project team, establish project governance, define priorities, review the scope and transition, and
from there create a project plan. Your partner will be doing the same,
complementary to your effort but not always in lock step with you. A few
words about that in a moment.

Partners will vary in how much they will drive this process. At IBM,
we have our own best practices that have proven successful for thousands of cloud implementations, which is why our recommendation is to
onboard with us as early in the project process as possible.

Tip! It is important to remember that while you are onboarding your partner, your partner is onboarding you. Every cloud
management partner handles it differently, so it is important
from the beginning to scope out that relationship and how
communication will happen and how often.

A cloud-managed service partner is different from many other partners you may have had for other projects in that you are not only creating and launching a technology solution, but in this case the relationship
will extend well past the transition into the future. With an SAP cloud
implementation, many customers find that a great burden of maintenance
and updates has been lifted from their respective shoulders, so it also will
result in a significant change in their staffing and where customers put
their technology resources. Customers usually shift from management
of technology to focus their resources on their core competencies, drive
value to their business, and increase revenue.

As we mentioned earlier, there is also some varying perspective by
your cloud management partner regarding the steps you will take. At IBM,

our cloud-managed services for SAP applications encompass a continuum of SAP project and application support services from the very beginning of onboarding with the customer onto the cloud environment, through go-live, and well beyond. In our case, the partner takes a larger role in the SAP project development phase into the SAP application production phase. We have a transformation service team that supports establishing a new SAP environment on the cloud infrastructure and, if applicable, migrating existing SAP environments to the cloud infrastructure.

So while your cloud implementation will follow the standard four phases of project management discipline (Launch, Model, Build, and Transition), your partner may view it somewhat differently. It is important that you are clear from the beginning with your partner regarding phases, vocabulary, and more as this will become important in establishing milestones for your project.

At IBM, for example, we have a transformation services team that implements the plans, processes, and tools necessary to transform the SAP application environment to the steady state architecture, with all the necessary processes and service-level agreements (SLAs). We tend to look at the project as a three-phase process rather than four, emulating standard Information Technology Infrastructure Library (ITIL) methodology. As such, we call our phases "Onboarding," "Build and Delivery," and "Transition." From our perspective, the first part of the Onboarding phase, designed to establish the foundation for your cloud-managed services onto the SAP applications cloud environment, includes the following steps from what is more traditionally referred to as the Launch phase.

- Create a detailed project plan and identify critical milestones and dependencies.

- Establish the project organization charts, including defining roles and responsibilities and identifying project sponsors and the quality assurance (QA) team.

- Establish the project status reporting and governance processes.

- Launch the project team, including reviewing project scope, project milestones, priorities, and roles and responsibilities.

In Chapter 5, we will go over the remaining steps in the Onboarding phase that are in what is often called the Model phase.

CHAPTER 4

Step One: Assemble a Project Team

First, let's assume you've properly embarked on a project. You began with a directive from the executive team out of which has emerged an executive sponsor for the project who champions it throughout its life cycle. That executive sponsor may or may not come out of the business, but at the very least you have demonstrable business support for your project.

You've done your due diligence by building a business case for a cloud implementation; now it's time to align your project team, as **Figure 4.2** demonstrates.

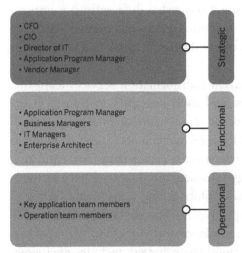

Figure 4.2 Align your project team into the three over-arching pursuits: strategic, functional, and operational

Your company's executive technology team is on board: CIO, CFO, director of IT. Your application program manager, vendor manager, business manager, IT manager, and enterprise architects are in place and understand their roles in the project. Your key application and operation team members have been consulted and understand that they will have a role in this effort as it hurtles down the track.

We recommend that our clients all have an SAP architect on staff who is responsible for the strategic components of the SAP environment including release planning, capacity planning, environment planning, and the technical integration management aspects of the SAP environments.

The client SAP architect also owns the SAP technical procedures and is a key stakeholder in designing them (with the assistance of the partner) and approving them.

About Finding and Working with a Partner

According to a Frost & Sullivan survey, 91 percent of businesses have or will use a partner or other third-party expert to implement their cloud strategy. The same firm found that 52 percent of enterprises report they have insufficient cloud expertise on their staff.

For most organizations, it is not so much if, but *when* they will find a partner for their cloud-based initiatives. Partners come in all varieties but there are several metrics by which to measure a partner, be it obtaining technical assistance to make your own move to the cloud or a full-service managed cloud provider. Some things to look for in a managed services provider include the following:

- **Cloud infrastructure.** Does it have data centers in proximity to the markets you wish to serve? For example, global companies know that many countries require the data generated from its business be stored in that country. This is an important consideration before selecting a partner who may or may not have a presence in that country.

- **Library of infrastructure and application services.** Many cloud partners have a limited library.

- **Standards-based, automated processes.** Surprisingly, this is not always the norm among cloud providers.

- **Infrastructure options flexible enough to handle current and future needs.** Does your cloud partner have the ability to provide a private cloud, if that is required, or a shared cloud?

- **Service-Level Agreements (SLAs).** This may sound counter-intuitive, but one of the initial discussions you need to have with any potential partner is its SLAs.

- **Security.** We're not only talking about SAP-level security, but the physical security of the data centers. Who has access?

- **Expertise.** When it comes to the cloud, expertise is probably your number one requirement from a cloud vendor.

Expertise in particular covers a lot of areas. Certainly, any partner you select needs to have application-specific expertise, but also knowledge and experience with the applications you may require in the future. Your partner should be able to provide concrete advice for assessing and selecting applications for the cloud. While you can always split your cloud management needs between vendors, you lose a significant amount of the efficiency you gain by hosting your SAP applications in the same data centers.

Other areas of expertise you should consider are core expertise in infrastructure, workload, application management, and professional services. Your partner also should have experience planning and preparing your applications for the move, and then experience in migrating those same applications to the cloud. You need to make certain that the partner who manages the hosting has policies and procedures and, most importantly, a track record of continuously optimizing the cloud. It also must offer compliance services that meet rigorous standards for privacy and security.

It should go without saying that whoever you partner with must have depth of experience with SAP on the cloud.

Many SAP customers seeking a partner require flexibility, such as a choice of deployment options and services. Many want a unified platform experience, with scalable computing selections. Others seek speed to market so a full provisioned environment, ready to go, is desired, with proper OSs, security patching, installed middleware, and fully configured virtual machines (VMs).

Your Partner's Project Team

Your partner also must have a project team as shown in **Figure 4.3**.

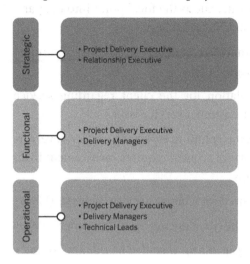

Figure 4.3 Your cloud partner's project should align with the strategic, functional, and operational buckets

As with all teams, it needs to have someone at the head. Let's call that person the Project Delivery Executive.

Project Delivery Executive

The project delivery executive makes the transition and implementation phases go smoothly. This is your main service delivery liaison with your partner and the SAP solution. As the escalation point for critical business activities, the project delivery executive is responsible for achieving customer satisfaction, so take advantage of this valuable resource. The Project Delivery Executive performs the following specific tasks:

- Provides program management for projects during the contract-to-production and post-production projects phases

- Trains users on the management portal tools and on the Cloud Managed Services for SAP Applications processes and procedures

- Manages problems and escalations as necessary

- Holds regular operational review meetings to review service requests, service levels, and features

- Serves as a customer advocate, and is the focal point into the partner and, by extension, Cloud Managed Services

- Is a contract expert and therefore manages contract communication, clarification, and disagreements

- Acts as the point of escalation for the client regarding service requests, delivery, or other customer service issues

- Coordinates service request resolution, especially for first-level service requests and when service request requires involvement from multiple performing groups

- Obtains downtime as required for Change Service Requests (CSRs)

- Communicates updates to the client during multi-customer outages

- Performs the project manager role for installed based non-complex projects

- Engages and updates the executive sponsor as required

Now don't panic if you've selected a partner and it doesn't have a role called a "Project Delivery Executive." What is important is that there is someone on the partner side to fulfill the same role.

Tip! It is always a best practice to have a single point of contact within your partner organization and an escalation path (up the chain, not laterally or downward) to respond to issues as they arise. Do not accept alternate points of contact that will not give you the attention your issue requires.

At IBM, for example, we split the job of project delivery executive among two roles:

- **A project executive (PE)**, the "relationship executive," who is responsible for all contract and financial aspects and is the main escalation point for the customer.

- **A delivery project executive**, the contact dedicated to the client who manages all day-to-day delivery aspects. This includes issue resolution, planning of tasks to fulfill new requirements, monthly status reporting, and more.

Other roles on the partner team should include the following:

Project Managers

Depending on complexity, your partner usually will assign one or more project managers to your project. The project manager assists with new implementations, upgrades, or complex migrations. The project manager focuses on managing the overall success of the project and coordinates the partner resources with the objective to complete the project. The project manager's performance should be tracked and measured against completing the project on time, in scope, on budget, and with high quality to ensure the manager has the same vested interest in the success of the project as you do.

Primary Delivery Technical Lead

During the production management phase, the partner will assign a dedicated technical lead as your lead technical support engineer. Your technical lead is responsible for the overall resolution of technical issues with your applications, and coordinates with other technical teams within the partner's delivery organization to resolve issues as necessary. The technical lead performs the following tasks:

- Reviews all service requests with you and the project delivery executive on a regular basis
- Manages all first-level calls
- Develops all first-level reviews
- Participates in regular operational review meetings
- Manages all changes to your system's architecture

Establish Project Governance

Project charters are an effective way to formally kick off a project, albeit in our experience not a lot of companies do this. One of the challenges is that projects have a tendency to evolve and as the customer does more research and investigation, scopes shrink or grow, high-level requirements change, and budgets are increased or decreased.

However, the advantage of creating a charter at the beginning is that it puts a stake in the ground for stakeholders, a snapshot moment with which support can be gained for the project internally, especially among the owners of the business processes involved. Also, it is a living document that can be referred to throughout the life of the project and updated, or better, serve as the launch document from which the metrics of success can be established and measured.

One of the reasons we recommend getting a partner on board as early as possible is to help draft a charter document that reflects the new world of the cloud that your company is moving into. Most companies we work with have limited experience porting SAP applications on the cloud, and, as a result, they begin with assumptions that in many cases are not tenable or sometimes even not realistic.

The charter identifies the stakeholders of the project, not only within IT, but in the business, partner, and supply chain communities. It outlines, in high-level terms, the resources that can be made available as well as the steps to obtain those resources. It identifies in broad strokes the responsibilities required, and as well contains a discussion of the risks, constraints, and any assumptions that can be attributed either to the current state or to the project. A project charter or whatever your project management culture calls it should have the following facets:

- Project sponsor
- Project manager or program manager
- A high-level description of the project
- The purpose of the project, as defined by the three axes: increase revenue, reduce expense, mitigate risk
- High-level objectives on how the project purpose will be defined as a success

- Requirements, at a high level

- Anticipated budget

- Project schedule expectations, with some indication of milestones

- Expected project outcomes. Not only should you consider the ROI and the three axes, but also those areas that are unique to the cloud, such as flexible provisioning or expansion of customer engagement.

Tip! Not only should the project charter or equivalent be developed in advance of the kick-off meeting, it should have been reviewed and signed off by internal stakeholders. The kick-off meeting should not be the event at which the charter is discussed and debated.

Organization/Meeting Chart

Another task to complete in advance of the kick-off meeting is the project organizational chart with an outline of a proposed meeting schedule. One reason to combine the two into one is that it becomes more intuitive to show the relationship of the staffing of the project between the customer and partner.

Our approach is to break the dynamics of the project into three areas, strategic, functional, and operational. While participants of the project may overlap among those areas (and should to ensure full transparency and shared communication channels), separating those channels moves the project along without bogging down all the participants in interminable and pointless meetings.

Project Strategic Level

At the project's highest executive level, there should be occasional meetings, but in no way should this be a limitation, because in many instances this level will meet more often when significant issues have been escalated.

This is the arena in which the customer's executive level, headed by the executive sponsor, executive stakeholders, the program/project manager, and those responsible for the relationship with the partner, meet with the partner's project executive and, depending on the project and make-up of the partner, the relationship executive.

The purpose of the strategic meetings is to review and refine the contracted relationship between the parties. It also has the purpose of ensuring that the strategic objectives behind the project are in alignment with the objectives of the operational business role within an organization. As high-level requirements and plans change, this should be the group to evaluate the overall direction of the project to make certain it not only has adhered to the previous strategy but that it can also effectively move to incorporate any new strategies that have evolved during the course of the project.

One of the most important purposes behind this group is that it serves as a sounding board for the customer's roadmap for future innovation on the cloud. While the customer can share its vision, it is also an appropriate venue for the cloud partner to make the customer aware of innovations within the industry as a whole.

Project Functional Level

As you can see in **Figure 4.4**, the functional team meets monthly (three times more often than the strategic level), but not as often as those responsible for the day-to-day progress of the project.

At the functional level, the customer's program/project manager and relevant business managers, IT managers, and enterprise architect, sit down with the partner's project executive and project delivery executive (and delivery managers, if such a role exists) to review the overall performance of the project against the metrics set at the outset. This group also reviews any issues, risks, or opportunities that have arisen during the course of the project, always with an eye on determining what, if any, should be escalated. This is also the level at which the focus is on executing the strategy set from the strategic level, focusing on service delivery and the technology direction, as well as implementing the innovation roadmap.

Project Operational Level

This is the group responsible for the day-to-day execution of the project. As seen in Figure 4.4, it meets weekly, which is four-to-five times more often than the functional level, and 12 times more often than the strategic level.

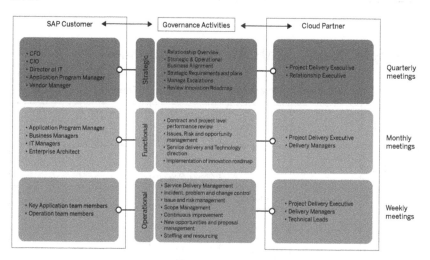

Figure 4.4 Once the project teams are aligned, each area proceeds along its own timeline and own activities

From the customer side, the program/project manager is joined by key application and key operations personnel, which together is the team that performs the hands-on execution of the cloud project. On the part-ner side, the project delivery executive brings along the delivery managers and technical leads.

This level wrestles with service delivery management, ensuring the platform containing the client's dedicated SAP application environment is being constructed and operating within the parameters established. This group also maintains and executes against the tracking of incidents, issues, and change control, making certain that the work to resolve or address those items is being addressed. This also is the area where scope is carefully monitored and, when "scope creep" appears, is escalated. These individuals also are responsible for continuous improvement and track-ing that metric throughout the project.

While Figure 4.4 is a generic view, it is critical that these roles correspond to the names of actual individuals in addition to identifying the escalation path within the organization. Everyone involved with the project should have access to the organization chart and escalation path. See **Figure 4.5** for an example of a typical project meeting schedule.

Example Steady State Meeting Cadence

	Monday	Tuesday	Wednesday	Thursday	Friday
Week 1			Weekly Operational & Change Meeting		
Week 2			Weekly Operational & Change Meeting		
Week 3		Monthly PMO Meeting	Weekly Operational & Change Meeting		
Week 4		Quality Steering Committee Meeting	Weekly Operational & Change Meeting		

Executive Steering | Program Management | Operational Oversight | Operational Management

Figure 4.5 A monthly view of a typical project meeting schedule

Kick-Off Meeting

The "meeting of the minds" of the project occurs at the kick-off meeting. This is often the first time everyone—executive sponsor, project leads for you and for your partners, and stakeholders—will meet to agree and commit to how the project will proceed. Some of the objectives to be covered in the kick-off meeting should include the following:

Review the Transition and Scope

In ITIL terms, this is referred to as the Process Objective, the purpose of which is to make certain that the service transitions are in line with the organization's project management guidelines, policies, and procedures.

Within an SAP cloud project, due to the significant role of the partner, this also should be the step in which the service transitions are outlined in a way that is complementary with the project management guidelines of the partner as well. At IBM, for example, we are well known for having our own vocabulary and processes, all developed as a result of thousands of migrations by clients all over the world. Therefore, this step should be viewed as a point at which the SAP customer and the cloud-managed service provider achieve a meeting of the minds regarding what each means by the various phases, steps, and terms used in project planning.

It is important to note that this is not the point at which the project plan is hashed out in detail, but instead looks at the planning from a high level to provide direction for the plan when it is created.

This is also the starting point for which the scope is defined and codified. The kick-off is an excellent opportunity for a back and forth between stakeholders, the IT team, and the partner to achieve the meeting of the minds regarding the objectives. Equally important, it is an opportunity to define the limits of the parameters of the project, as well as to explain the impact on the project to budget, time, and risk based on changes to the overall scope.

Define Priorities

The kick-off meeting is also a perfect time to discuss broad priorities of the customer throughout the various phases of the project. Priorities vary and priorities change, but codifying them at the very beginning of the project creates a baseline of discussion that can reviewed and readdressed throughout the project at the strategic level. Some areas of priority to look at include the following:

- Protecting revenue
- Uptime and/or disruption of mission-critical systems
- Patch management
- Security of data
- Performance

Refine Roles and Responsibilities

The kick-off is the ideal time, with everyone in the room, to spell out roles and responsibilities, but one of the major side activities is also to establish rules of engagement. Lines of escalation and communication channels are frequently overlooked or minimized in the process of creating a project.

Who has responsibility for project control? That is, who monitors its progress on the customer and partner side, and what is the process of reconciliation between those two narratives? And when there is conflict, how is it resolved?

Who is responsible for project reporting? Where is the repository or repositories? Also, once communication channels are identified, who has responsibility for managing them and maintaining them? Who is responsible for communicating what to whom? How are documents ratified, for example, who must sign off on the project plan? Where is it maintained and updated? What is the process for changes?

This is all basic project management and should come as little to no surprise to anyone, but in our experience if these roles and responsibilities are not detailed or explained at that kick-off, there will never be another opportunity when all executive team, stakeholders, project team, and partner team will be together in the same room.

The speed and cost of any move to the cloud is determined by three factors: the complexity of the landscape, the size of the database to be migrated, and the acceptable level of business downtime your organization can tolerate.

With regard to complexity, we are assuming that there is some SAP or ERP-related landscape that you want to move. If SAP, is it a current, supported version? Are you on Unicode, which is a must for the cloud? Depending on that complexity, you may require regression testing as you proceed.

The size of the database, by comparison, is in part self-explanatory. The bigger the database, the longer the migration. What you are migrating the database to also plays a significant role. For example, if you are going from Financials in SAP 4.6 to SAP S/4HANA Finance with a large number of feature changes, you will find the move a complex one.

Your business downtime window also presents a challenge that only you and those who manage your business can determine. With all the time in the world, the complexity and size of your database pose no challenge. When the window is limited, it means more tools need to brought in to make the transition. More tools mean more cost for the migration.

In the next phase, the "Model" phase, it is important to understand any potential changes required in the integration points, business processes, and system customizations. Conducting an SAP HANA upgrade and migration runs in a sandbox environment prior to finalizing the plan can be a great approach to identifying requirements in all those areas, as well as any cloud infrastructure requirements.

The objective here is to get the migration strategy right. Everyone needs to be involved; technical team, application support team for remediation, and super users for go-no/go verifications. You also need to develop a back-out plan as a contingency should the worst-case scenario occur. It is rare, but without proper consideration, if it does happen the result can be disastrous.

Getting back to our example company we introduced earlier:

One of the first steps following onboarding is related to validation and planning, in this case we conducted a "Workload/Network Compliance Discovery." To be able to define the perfect migration path it was important to gather key data on the existing environment: servers, workloads, operating systems, application, data, and data dependencies between applications, to name just a few examples. We used data collection tools to complete this phase as quickly as possible. After collection, all data has to be analyzed and assessed—and sorted by certain criteria.

Here it is important that the end state is considered when planning the migrations. Since the SAP application architecture footprint consists of multiple SAP applications, it is important to understand which ones are required to move to SAP HANA and which ones can move to the Adaptive Server Enterprise (ASE) database. Note: Although some SAP customers migrate all applications to SAP HANA, many look at the SAP HANA license costs and the potential improvements in business processing before deciding on the SAP HANA strategy from an application perspective.

68

Phase Two: Model

Onboard to The Cloud, Part Two

Now we come to everyone's favorite part of any project: documentation. For example, now that you've created a project team and defined the escalation paths with your partner, you need to make certain that information is a deliverable within the project plan.

Before the project can kick into high gear, requirements must be defined, SAP and other software licenses obtained, workload analyzed, IPs identified, and much, much more. The more complete you can be in your initial documentation phase, the faster the project will move and the fewer complications you will experience. This is even truer in an SAP cloud project than most other projects since so much is dependent upon clear and concise information with your cloud partner.

Before you even write your project plan, there is a tremendous amount of information your cloud partner requires. What follows in this chapter are examples of the type of information sought and approaches that we use at IBM, but they should be applicable to most, if not all of your potential cloud partners. While much of this may seem overwhelming, we will explain each information request in the next chapter.

Provisioning Background Documentation

Your cloud managed service (CMS) provider, in order to efficiently set up your compute instances, requires an extensive amount of information from you regarding your current landscape and the network components that will eventually connect you to your cloud instance (more on this later). It may seem overwhelming to some, but the more detail you can include, the better. At IBM, for example, we ask for the following information. (Note that along with each item, we have identified the role in your organization where many times this information can be found):

- Current database information — Customer/Solution Architect
- Virtual machine (VM) name — OS Support
- Host name — Customer/Solution Architect
- Server name(s) — Customer/Solution Architect
- OS — Customer/Solution Architect
- RAM (GB) — Customer/Solution Architect
- Number of CPUs — Customer/Solution Architect
- Management service — Customer/Solution Architect
- VM (non-SAP HANA) storage (in gigabytes) per server — Customer/Solution Architect
- SAP HANA storage (GB) per server — Customer/Solution Architect
- Server type — Customer/Solution Architect
- Datacenter — Customer/Solution Architect
- Landscape tier — Customer/Solution Architect
- Phase number — Customer/Solution Architect
- Instance number — Customer/Solution Architect
- System identifier (SID) — Customer/Solution Architect
- Application Release/Patch (Version Release) — Customer/Solution Architect (required if you chose "Fully Managed" or "IaaS Plus and DB" in "Management Service")
- Database — Customer/Solution Architect (Required if you chose "Fully Managed" or "IaaS Plus and DB" in "Management Service")
- Database version — Customer/Solution Architect (Required if you chose "Fully Managed" or "IaaS Plus and DB" in "Management Service")

- Database encrypted? (Y/N?) — Customer/Solution Architect (Required if you chose "Fully Managed" or "IaaS Plus and DB" in "Management Service")
- Domain name — Customer/Solution Architect
- Virtual Local Area Network (VLAN) ID — Network
- NTP Server IP — Customer/Solution Architect
- Time zone — Customer/Solution Architect
- IZ or DMZ — Network
- Public IP — Network
- Server IP — OS Support
- Subnet Mask — OS Support
- Gateway — OS Support
- CUSTOMER NTP SVR — Customer/Solution Architect (Required if you chose "Fully Managed" or "IaaS Plus and DB" in "Management Service")
- DNS Search Domain Name — Customer/Solution Architect (Required if you chose "Fully Managed" or "IaaS Plus and DB" in "Management Service")
- Interface Static Route — Network (Required if you chose "Fully Managed" or "IaaS Plus and DB" in "Management Service")
- Customer DNS Server 1 — Customer/Solution Architect
- Customer DNS Server 2 — Customer/Solution Architect
- IMZ VLAN ID — Network
- IMZ IP address — Network
- IMZ Subnet Mask — Network
- IMZ Gateway — Network
- IMZ Routes — Network
- SL IP address — Network
- SL Subnet Mask — Network

CHAPTER 5

- SL Gateway — Network
- SL Routes — Network
- Language — Customer/Solution Architect (English is default)

Network Documentation

You need to identify the path for how you will connect to the VLAN created by your cloud partner. While we discuss the details later, you will be asked to provide:

- Subnet purpose
- Customer (CNS) IP range
- Customer (CNS) IP gateway
- Customer (CNS) VLAN
- Cloud partner management (IMZ) IP range
- Cloud partner management (IMZ) gateway
- Cloud partner management (IMZ) VLAN
- Route statement 1
- Route Statement 2

You also will be asked to provide your subnet ranges, which in turn will be matched by the IBM Management Zone (IMZ) subnet assigned by your cloud partner. **Figure 5.1** is an example of the type of documentation on network connections that your partner may ask you to provide. There will be much more information on network connections in the next chapter.

Domain Name System (DNS) Information

Customer DNS Information		
Host Name	Domain Name	Server IP
Details		
DNS Solution		
DNS IP Addresses		
Domain		

Figure 5.1 Example of a documentation form related to network connections you may be asked to fill out and return to your cloud partner

Parameters for the VPN Tunnel

Like many cloud management services, we use an Internet Protocol Security (IPsec) VPN tunnel for the end-to-end communication from your firewall to ours. Whatever method of communication you use, you will be asked to provide information such as what appears in **Figure 5.2**. Note that while certain key management parameters must match, the values attributed to those parameters are flexible. We've published the defaults here, but your team can apply its own values.

VPN Encryption Parameters		Confirm or change the default values in this column
Key Management Parameters	Required to Match	Value (default)
Key Management Scheme	Yes	IKEv1 Main Mode
Encryption Algorithm	Yes	AES128(preferred) or 3DES
Hashing Algorithm	Yes	SHA-1 (preferred) or MD5
Authentication Method	Yes	Pre-Shared-Key – Generated by HEC, shared out-of-band
Diffe-Hellman Group	Yes	Group 2
IKE Renegotiation Interval	No	24 hours (86400 sec)
ISAKMP Keepalive Interval	No	10 Seconds, 2 Second Retry

Phase 2 VPN Encryption Parameters

IPSec Parameters	Required to Match	Value (default)
Protocol	Yes	ESP
Encryption Algorithm	Yes	AES128(preferred) or 3DES
Data Integrity Algorithm	Yes	SHA-1 (preferred) or MD5
PFS (Perfect Forward Secrecy)	Yes	Enabled(preferred)
IPSec SA Renegotiation Interval	Yes	4500MB/3600s

Figure 5.2 Example of VPN Parameters form your cloud partner may request of you

Firewall Rules

Your cloud partner will need access to your systems, and therefore will need to know the rules to your firewall. Is it unidirectional, one way, where the source to destination traffic can only be negotiated from the source? Is it bidirectional, where traffic is negotiated between source and destination from both ways?

We generally send our clients a form seeking the source network IP address translation (SNAT). Traffic in packets is sent by the device with the private address (e.g., 192.168.1.6) behind the firewall to a public IP address (e.g. 172.20.44.55), as well as the reverse, meaning the destination network IP address translation (DNAT) and the public IP address (e.g. 172.20.44.55) from which packets are sent from outside the firewall and translated to the private address (e.g. 192.168.1.6) behind the firewall.

Note that the network address translation (NAT) is a 1:1 IP translation and requires more public IP addresses when listening services run on different private IP addresses. The port address translation or network address port translation (PAT/NAPT) is a one-to-many IP PORTS translation. One public IP address can be used for multiple services on different private IP addresses.

Your cloud partner will need your SAP security information, such as what is requested in **Figure 5.3**.

General Information	
SAP User ID	
SAP Password	
SAP Applications List - To be Built	BW/BPC, ECC, etc
Configuration Client #	

Figure 5.3 Your cloud partner will require your SAP credentials; this is an example of the type of form your cloud partner may send to you

Key Dates

To help set expectations about what progress is expected for the project (or projects), we generally ask our clients to provide a broad overview of what they expect when. **Figure 5.4** is a fantastic starting point for building a preliminary, high-level project plan and also will help you and your cloud partner to come to terms on what eventually will be the completion criteria for the project.

Customer Name:	Phase 1 Name	Phase 2	Phase 3	Phase 4
Environments (i.e. - Dev, QA, PRD) in Phase				
SAP Products				
# of SAP Instances				
# of VMs				
# of Physical Machines				
POD Location				
DR requested (y/n) *Note commit dates do not apply				
Requested Delivery Date of each phase				
Committed Delivery Dates by Phase				
Critical dates:				
Commit expiration date (close by date)				
Handoff to Delivery by				
Start Onboarding				
VST PO Received by				
SAP HANA IN PLACE				

Figure 5.4 Form for identifying and specifying project milestones

The Project Plan

At IBM, we have done so many cloud implementations that sometimes it seems as if our project plan templates have templates. Most of our projects related to our cloud managed services follow certain steps. Of course, not every project is the same, but most projects follow a very similar path. What follows is an overview of our standard project progression, which should help guide your cloud project.

Pre-Enablement Phase

This is the term IBM uses for the initial steps in the Launch phase, and Chapter 4 walked you through many of the steps. Here's a brief summary of the milestones:

Sponsor the project

This is the beginning of the project at the highest levels, where the customer identifies an opportunity to increase revenue, reduce expenses, or mitigate risk.

Identify the stakeholders

For every project there are stakeholders, both internally (IT and executives in charge of a line-of-business) and many times externally (key customer groups and supply chain partners). This step pinpoints where support for the project will be found and who will be participating in the planning and execution.

Define the project

This step varies based on corporate culture, as sometimes it is restricted to just a particular phase or phases. The heart is to understand the objectives of the project and outline the activities to achieve success.

Deliver transition work items/deliverables

This is the handoff of all the above activities to your cloud partner or integrator.

Conduct due diligence

We always include this as an optional step. This is where you vet the steps so far and determine if the stakeholders, cloud partner, and potential project team are an appropriate fit to achieve the objectives of the project. This milestone many times is where gaps and risks are identified and issues noted, as well the accompanying proposed solutions to the same.

Create a high-level project plan

This step is more detailed than a back-of-the-envelope guess at milestones and delivery dates. Our hope is that thanks to books such as this one, as well as discussions with the potential project team and your partner, you can make a fairly accurate accounting of the milestones and of the delivery dates.

CHAPTER 5

Define the project governance

While also part of "Launch," we put it after the kick-off in the project plan but not dependent upon it. Tasks in the project governance milestone including locking down the project scope, scheduling the meeting calendar, creating the communications plan as it relates to governance and escalation, and working out the exchange of status reports.

Hold the kick-off meeting

From our perspective, this is where the Launch phase concludes and the real work begins.

Transformation Planning Phase

If this phase were on a map, you would see a red star that says, "You are here." In this phase, you will gather requirements and model your new implementation.

Build the network and switches

While it may seem unintuitive to build the connection before building anything else, so many later tasks can be dependent upon the construction of a network, switches, and connective software that we always make it a priority at the beginning. Think of this connection as the umbilical cord for your cloud implementation.

Design and implement Network Attached Storage (NAS)

If the project is determined to require NAS, this step will define who acquires the storage and where it will be located.

Create the project plan

Perhaps the project plan milestone is the most self-explanatory step of all in the project plan.

There also are several steps related to validating the information you provided, which was detailed above, namely firewall rules, software licenses, and more.

Execution Phase

This is the "Build" phase of the project.

Build DEV OS/build the DEV application/build the QA OS/build the QA app

These tasks are for your cloud partner, getting the development infrastructure ready for your implementation.

Testing

Once the DEV system is in place, work begins on building and testing your new cloud environment. Upon approval, signed off by you, your cloud partner will repeat the Build steps for the production environment, followed by more customer testing and signoff.

Rehearse a mock cutover

The final step before go-live is to conduct the mock cutover. the purpose of which is to identify any potential issues before moving to the new production system.

Production/Go-Live Phase

This and the next phase represent the Transition part of the project.

Go-live

This is another self-explanatory milestone, one that concludes with your acceptance.

Balance loads

Now that the system is in production, load balancing can be tuned.

Implement printers

If a printing infrastructure is required, now is the time.

CHAPTER 5

Preparation to Final State Phase

Create the final documentation

Although your cloud-managed service will be hosted and maintained by your partner, that is still no reason not to make sure you come away with proper documentation for your SAP and related systems.

Achieve customer acceptance

For this milestone, the contract and agreements are reviewed to make certain all the terms and conditions of the project have been met. Your cloud partner also should conduct an OS and application QA validation. This also begins the process of monitoring service-level agreements (SLAs) and other ongoing policies and procedures.

Close the project

This milestone not only formally closes the project, it also is where you and your implementation team take a look back and identify lessons learned. This step, which many organizations never seem to complete, is probably more important for a cloud implementation than any other project. The rationale is that this will be the first of many cloud implementations and the lessons learned with this project in many cases are directly attributable to future projects. Don't skip it!

Requirements Gathering via Workload Analysis

Every organization has (or should have) its own methodology for generating requirements for a project. Because so many cloud implementations are migrating existing SAP instances onto the cloud, either in whole or in part, we have developed an approach to analyzing your current workload state and then using that analysis to extrapolate the requirements for your future state.

"Workload" is the aggregate of the requests made upon a computing system by users and applications. It describes the requests run against a set of servers, OSs, and application products with the business code on top of that. A business service is a series of interrelated workloads required to complete a specific task or tasks.

Workload Analysis Digs Deeper

Workload analysis covers four different dimensions (see **Figure 5.5**):

- **Workload characteristic fit analysis.** This represents the qualitative characteristics of the application related to the infrastructure.

- **Quantitative workload fit analysis.** This represents a data inventory of the servers to pinpoint OS, CPU, memory, and any other functional requirements.

- **Workload assessment analysis.** This assessment evaluates your workload at the application layer along with its underpinnings, such as the development technologies that they are using. You need to identify the functions you are using, determine how you will develop applications, and what your release approach will be.

- **Workload affinity mapping analysis.** Once you've done all of this, you finish by identifying how all these applications and their commensurate workloads interact with each other. Just like the days of data center migrations, you can't take one application and move it to a new environment without understanding the ramifications of its interactions with other applications. You need a mechanism to assess and quantify that impact so you can make a fully informed decision about cloud infrastructure.

CHAPTER 5

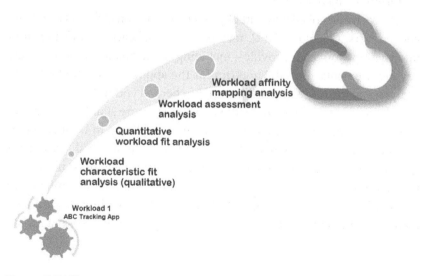

Workload affinity
mapping analysis

Workload assessment
analysis

Quantitative
workload fit analysis

Workload
characteristic fit
analysis (qualitative)

Workload 1
ABC Tracking App

Figure 5.5 When assessing existing workloads for the cloud, a workload point of view must consider four key dimensions

Let's examine each of these steps in greater detail.

Workload Characteristic Fit Analysis

When considering a cloud infrastructure, you need to test the business application or workload against the following:

- What are the scalability requirements for the workload, both for now and in the future?
- How important is speed-of-deployment for this workload?
- Which better serves the needs of the business, an application or software-as-a-service (SaaS) on the cloud?
- How differentiated is the workload from on-premise (if it currently exists) versus the cloud state and, by extension, is it a source of competitive advantage?

Answers to these questions will guide you to the cloud deployment model that best suits your needs.

Quantitative Workload Fit Analysis

This level of analysis involves digging into your current environment and performing data discovery.

Above we described how part of your documentation at the outset is to list your servers, OS types, CPU, memory, and applications. But the more information you can provide—including performance metrics, security requirements, and SLA metrics for the applications—the better the analysis and the more dimensions can be checked and tested.

The goal is to understand from an infrastructure standpoint how the applications running on this infrastructure would map to what is offered via potential cloud providers. The greater the granularity, the better the recommendation as to whether the business service will fit or not (a true yes-or-no proposition). If it does fit, how much will it cost to migrate it? What will be the level of difficulty to do so? By providing performance metrics your cloud provider can better identify areas for cost savings.

Workload Assessment Analysis

Figure 5.6 highlights the underpinnings of a business service. By mapping your own systems, you can identify redundancies that can be consolidated. Can you replace or the functionality you require on another, more efficient cloud platform? Or can it be rebuilt in the cloud at an affordable cost?

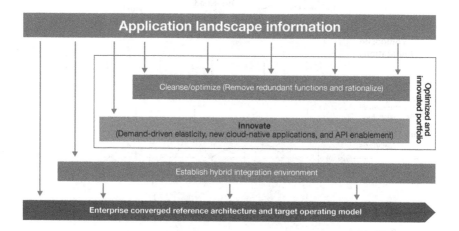

Figure 5.6 Use a workload point of view to build and manage an optimized and innovative portfolio with evolving hybrid IT architecture

Once this study is completed, the results need to be combined with your analysis of your infrastructure to determine the best path to the web. For example, some applications may be an easy fit to port onto your potential web partners based on your infrastructure review. However, when looking at the application layer, it may be more advantageous for the business to spend the money and time to rewrite the software in a cloud-native development technology.

Doing the study from an infrastructure *and* an application development perspective provides the full spectrum to make the appropriate decision for your needs.

Workload Affinity Mapping Analysis

You will need to build a map of workload integration points (**see Figure 5.7**). Of the workloads that you have assessed, do any have any critical integrations where, once moved to the cloud, they will create a major performance issue?

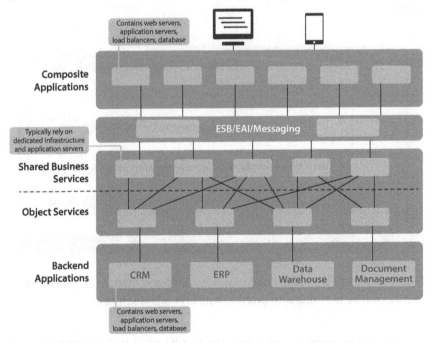

Figure 5.7 Don't forget to consider workload integration points

Application and Business Process Discovery

Because not all applications are appropriate for the cloud, sometimes there's very little benefit to be gained by moving a workload. Understanding what makes a workload cloud ready is important.

Finding those applications that are good candidates for cloud adoption is a key step here. At IBM, we developed a methodology that has been reliable at identifying these critical integration points for moving business services to the cloud: Once you have a workload that is ready to move to the cloud, we meet with the business and portfolio leaders of the applications and discuss, using a ranking perspective, how important these applications are to the business.

What we find in this step is that, when you ask that seemingly general question to any group, most will say that the software they use—or that their business unit uses—is the most important. But in this exercise, we want to strategically determine, from a revenue generation standpoint, how important each application is.

Once we get some firm financial metrics behind the applications under consideration for a move to the cloud, we can determine a ranking, and from there sit down with each individual application's owners and support teams and developers. We'll take this opportunity to understand all the individual transactions of those applications. And we'll work out, step by step, how they flow through the infrastructure.

We then create a non-production environment and have a test user simulate that transaction. We capture all the network traffic on all the server components and pull that together by timestamp so we can generate a report of one transaction. We then eliminate all the extraneous activities on the server during the transaction. This gives us an isolated, network capture of a single transaction as it flows from server to server, millisecond by millisecond.

We'll put that transaction capture into a modeling tool that enables us to insert simulated latency at critical integration points between the application on the cloud and any on-premise components.

One of the side benefits of this exercise is that we find integration points that the customer was not aware existed. We've found business services that, unbeknownst to the customer, were talking to non-production databases or even making calls out to the cloud.

This brings us to one of the many side benefits of this cloud workload process: cloud discovery. We find many, if not most, customers are already on the cloud, but don't realize it. We always find that our customers' individual business units or employees have appropriated cloud services, usually one-off solutions, and incorporated them into their workflows.

It could be anything from cloud storage, where they are keeping documents, to using cloud for disaster recovery. On average, a company is using more than 800 different cloud-based services and most of those are outside the control and knowledge of IT. Typically, the decision to use the cloud has bypassed IT and its rigorous controls.

When companies are looking to make a jump to the cloud, it is an ideal opportunity to identify where you currently are in the cloud and, more importantly, where your business might be at risk because of the lack of oversight or consistent controls.

IBM, for example, has a partner that will go into a company and examine the perimeter and other server logs from any corporate data center out to the Internet. From that data, it will produce a detailed report that can identify, down to the user level, interactions with the cloud. Not only will it tell you if a single user is storing documents in the cloud, it will tell you how often the user stores a document, how large the document is, and what the source of that document is.

During the Build phase, much of the heavy lifting will be done by your cloud partner. For past SAP projects, this was probably the most grueling and challenging phase, but because this is now in the hands of your cloud partner, your role should be less stressful.

As Chapter 6 will explain, you will use this time to focus on your change management and to make certain that your use case scenarios of the business processes you modeled are ready for your selected users to test on the new system.

The Build phase is where the systems are provisioned, the application programs and data are migrated, and the systems go live in the cloud environment.

This phase includes all migrations starting with the development environment and ending with the production environment and any potential resiliency environments for business contingency and disaster recovery.

Here, it is important to take any learnings gained from the development and test environment migrations to update and further detail from the runbook, that is, the documented procedures and operational activities of the system. When migrating to the production environment the project plan must be as "waterproof" as possible. Although the common saying is that "production is always different," the supporting systems further down the pipeline are the best preparation for the production runbook for the core migration activities. Where production is different in most environments is in the integrations and interfaces. They are more active in the production system than in the development and test systems. The non-functioning of any of them requires much faster resolution than the same in a non-production system. This is also why it is important to practice the integration and interface verification tasks in the non-production environments to understand any parameter changes prior to going into production migrations. In production, you only get one chance.

Back to our example company. The Build phases for the SAP HANA database migration and the SAP S/4HANA Finance transformation are not totally uncoupled because they happen in close proximity from a time standpoint. In the overall plan, an SAP Adaptive Server Enterprise (ASE) database migration can be compared to the initial system setup for the system integration work in a greenfield implementation—there is a strong dependency. Of course, the SAP HANA and ASE database migrations in existing live systems are more complex and especially the production systems require a flawless migration.

CHAPTER 6

Phase Three: Build

During the Build phase, the components for the application will be built. The best part, obviously, is that you don't have to build them.

The Build phase, in many respects, is the least intensive part of your cloud project and the one that requires the fewest internal resources. The infrastructure is almost always the responsibility of someone else, namely your cloud partner. But as we pointed out in the previous chapter, connecting to that infrastructure is a major responsibility for your team, hence the extensive questionnaire that we reviewed in the previous chapter. In this chapter we will go into detail what your organization must "build" to prepare for your cloud implementation.

While the provisioning and installation of the target SAP systems are for the most part the responsibility of your partner, you have significant tasks when it comes to the components required for the transformation/migration activities.

What follows is an overview of the Build and Delivery phase, which includes the following performance steps and responsibilities:

- Your cloud partner provisions the target SAP system.

- You provide the business process test scripts to be used to validate the database transition, interface migration, and batch job processing.

- You provide resources to thoroughly test the migrated system using documented scripts.

- You provide the version of the SAP application, database, and any support components, along with a list of all currently applied application patches.

- It is up to your cloud partner to provide a security interface file transfer standard.

> **IMPORTANT:** If you have any custom application code for interfaces that will be part of the engagement with the security interface from the cloud partner, it will be up to you, in most cases, to test that code and rework it. This can result in unexpected delays, so it is important to confirm the status on your end as soon as possible.

- If required, you will be responsible for providing a file server to hold interface files awaiting transfer to the delivery location of your cloud managed services partner.

- It will be up to you to identify which SAP source systems will be used as the master copy for each application module to be migrated. This will be applicable for all migration passes to be performed.

- Your team exports the designated databases and provides the resulting export files/master copy of the environment to your cloud partner.

- Your cloud partner delivers the master copy of the environment to the target SAP systems.

- Your cloud partner provides an empty database and imports your export files.

- Your cloud partner configures the database and application parameters based on database size, system workload, and available system infrastructure.

- Finally, your cloud partner gathers migration timeline information based on a proof of concept migration to incorporate lessons learned into project planning.

> **IMPORTANT:** Prior to any proof of concept testing, you should submit a report of any existing functionality, including customizations and interfaces, that currently are not working or that are subject to current support calls in the existing system.

The Network

Before you can do any of the build, you need to connect to your cloud partner. In the previous chapter, we walked through the extensive documentation required during the onboarding phase, documentation that is designed to walk you through the interconnection process with your cloud partner.

What follows is a more detailed explanation of your responsibilities when it comes to the network. **Figure 6.1** gives you an illustration:

General connectivity diagram:

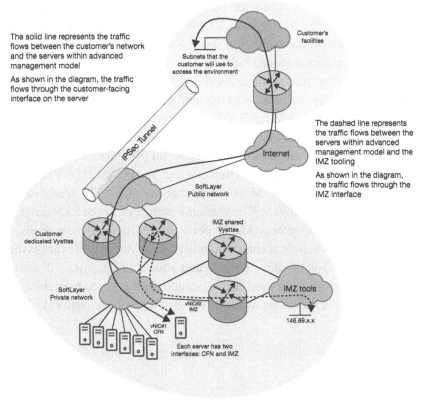

Figure 6.1 An overview of a networking scheme for a typical cloud implementation

To connect to the SAP on the cloud, you can have up to three customer-facing virtual local area networks (VLANs). You should consult with your cloud provider as to the number of VLANs needed to fulfill your requirements. Multiple VLANs provide a logical separation for the servers sitting on those VLANs, so if, for example, you require a production and a non-production landscape and want there to be no connectivity between the servers in those landscapes, you would need two VLANs to logically separate them. If, on the other hand, you want a production and non-production landscape but want connectivity between the two, one VLAN will suffice.

It will fall to you to provide one subnet, which we call a "customer facing network" (CFN), for each VLAN to be deployed. The size of the subnet depends on the number of servers on the VLAN, so make sure to dimension the subnet accordingly.

The advanced management module (AMM) within each server has two interfaces, one each for the CFN and another for the "management zone" (we call it the IMZ for "IBM management zone"). The CFN interface is the one that you will use to access the server. The IMZ interface is the one that the infrastructure and your cloud provider team will use to access the server. Every server therefore will require an IMZ VLAN for each CFN VLAN, and an IMZ subnet range for each CFN subnet range. The size of the IMZ range will be the same as the one of the CFN range in order to accommodate one IMZ IP per CFN IP. There are four available 16-bit ranges on each datacenter. You must choose one of those ranges that does not conflict with your network, and your cloud partner will assign the subnet portions to be used that is outside the range you have selected. **Figure 6.2** shows the ranges available for each domain controller (DC):

Customer's IMZ per DC				
Location Name	Customer Subnet	Customer Subnet 2	Customer Subnet 3	Customer Subnet 4
Dallas	10.4/16	10.68/16	10.132/16	10.196/16
London	10.5/16	10.69/16	10.133/16	10.197/16
Singapore	10.6/16	10.70/16	10.134/16	10.198/16
Frankfurt	10.7/16	10.71/16	10.135/16	10.199/16
Hong Kong	10.8/16	10.72/16	10.136/16	10.204/16
Toronto	10.9/16	10.73/16	10.137/16	10.205/16
Montreal	10.10/16	10.74/16	10.138/16	10.206/16
Wash., DC (WDC04)	10.12/16	10.76/16	10.140/16	10.208/16
Chennai, India	10.13/16	10.77/16	10.141/16	10.209/16

Figure 6.2 Customer IMZ per DC

The ranges that you select must not conflict with your network. They should be ranges that do not exist and will not exist in the future. **Figure 6.3** displays examples.

Subnet Purpose	Customer IP range	Customer IP gateway	Customer VLAN
PROD	10.25.0.0/24	10.25.0.1	
No-PROD	10.15.0.0/28	10.15.0.1	
Customer Provided Subnet	Enter Ranges Provided By Customers Below	IMZ Subnet	To Be Assigned By IBM Network Team
Considering Singapore	10.6/16		

Figure 6.3 Sample subnet ranges

Note that your IP gateway is always the first IP in the range, and your VLAN can't be defined until the onboarding process begins with your cloud partner.

VPNs

To set up the virtual private network (VPN), you will be required to define the parameters for the internet key exchange (IKE) phase one and phase two, with the exception of the pre-shared key (PSK) that your cloud partner will define and share with you during deployment. All parameters must match on both sides of the tunnel.

> **Tip!** We've seen several cases in which the cloud partner asks for a value and the customer uses a different value on its end. In some cases that causes the tunnel to come up, but it is or becomes unstable and causes connectivity issues and incidents. The parameters defined and agreed upon in this step are the ones that must be used to prevent potential issues.

The subnet or subnets that you defined in the previous step will be the range that your cloud provider will use when it defines the remote subnets for the encryption domain. Obviously if the wrong subnet range is listed, the cloud environment will be inaccessible. While this creates an issue, it

CHAPTER 6

is not unresolvable, but you need to make sure there is a process in place for escalating these types of issues.

Tip! Ranges such as 10.0.0.0/8 and 172.16.0.0/12 should be avoided if possible.

For each IPsec tunnel created for the new cloud implementation, you will need to provide the IP address of the device on your end that will serve as the termination point. In our questionnaire to customers, we ask for the address, all the corresponding network addresses, and the subnet mask.

Your cloud provider will provide the same on its side; however, that requires the deployment of a "virtual router." In IBM's case, we use Vyatta.

The process of setting up VPNs frequently is fraught with minor issues. It also is a specialty and while many are knowledgeable, not all are expert. One of the best pieces of advice we can offer is to make sure you have the best and most expert member of your team available for this step and have that person serve as the technical point of contact. After determining who will fulfill this role on your team, make certain the contact information is shared with your cloud partner. Many times issues arise that can be solved with a quick telephone call or even a text.

Firewalls

In the previous chapter, we explained in relative detail what information you need to provide to your cloud partner for it to navigate through your firewall(s).

By default, IBM and other cloud providers allow communication bidirectionally between your network and the cloud servers, but if your firewall in unidirectional, your cloud provider needs to know that.

From the cloud provider's side, we enable communication with the AMM over the following ports:

- ICMP
- TCP: 20-22, 25, 80, 443, 515, 1100-1399, 3000-3999, 4200-6999, 8000-8999, 9100, and 30000-65535
- UDP: 53

Sometimes additional ports are required and if so, this requires changes to the firewall rules, which for most cloud providers can be done at any time. However, we suggest again that you make certain there is an escalation path to make these kinds of changes.

Provisioning

Setting up the connection for provisioning will fall to your cloud partner, but there will be some tasks for which your team will be responsible. The CFN subnet, subnet mask each virtual machine's IP address will be defined by you and the solution architect.

Proof of Concept

We're not trying to confuse you, but there are "proofs of concept" and then there are "proofs of concept."

At IBM, we use the term to mean one of two things. We describe the first test of your cloud implementation as a "proof of concept" (more on that later). But another meaning has been referenced in an earlier chapter, in this case the "Entry Service Level" for SAP HANA on the cloud, which at IBM we have rebranded as the "Start-Up Bundle."

For products such as SAP S/4HANA Finance Cloud we offer the Start-Up Bundle to allow customers to test out the application on a managed cloud service in a way that does not require administrative support and is low risk, but at the same time is perfectly designed to go from a non-production development environment into a production environment. These products are pre-configured to make them as seamless and simple as possible as possible.

The Start-Up Bundle (see **Figure 6.4**) also is targeted for SAP landscapes for when you might need to run SAP systems for a short period of time (minimum three months) but don't require the full maintenance or services for a full production system. Typical scenarios include customers looking to pilot a new SAP application or to conduct modifications to existing systems in a proof-of-concept approach. They also are ideal as a training system or for just kicking the tires and exploring the capabilities of the applications.

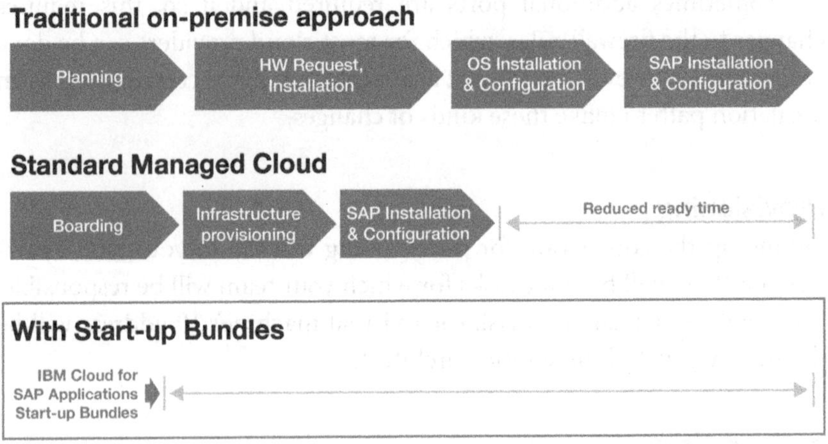

Traditional on-premise approach

| Planning | → | HW Request, Installation | → | OS Installation & Configuration | → | SAP Installation & Configuration |

Standard Managed Cloud

| Boarding | → | Infrastructure provisioning | → | SAP Installation & Configuration | ←———— Reduced ready time ————→ |

With Start-up Bundles

IBM Cloud for SAP Applications Start-up Bundles → ←—————————————————————————————→

Figure 6.4 The limited nature and very specific purpose of the Start-Up Bundles mean they can go live almost immediately

The Start-Up Bundles, however, are not intended to be used for significant project development or testing extending over a longer period of time. Like the Full and Development Service Class instances, IBM provisions servers, installs and configures databases and SAP application software, and provides SAP onboarding and migration services. Ongoing management and monitoring services are not provided for Start-Up Bundles because software components used to remotely monitor and manage Full and Development Service Class instances are not installed in Entry Service Class instances. Start-Up Bundle instances can be migrated into a Development Service or Full Service Class instance should you desire to make such a conversion. Table 6.1 lists the services provided with the Start-Up Bundle.

Start-Up Bundle	**Architecture Services:** SAP-certified architect validates configuration before deployment.
	Provisioning Services: Servers, storage, and network resources allocated to customer specification from cloud infrastructure.
	Infrastructure Configuration Services: Provisioned systems are configured with specific parameters and operating characteristics to support the specified SAP instances and landscapes.
	SAP - Database and Application Installation Services: Deployed according to IBM standard operating procedures and SAP installation guidelines.
	Application Installation Services: Deployed according to IBM standard operating procedures and SAP installation guidelines and documentation.
	Infrastructure Configuration Services: IBM configures operational infrastructure for the SAP landscape/instances.
	SAP Configuration Services: SAP and database applications are configured according to SAP installation guidelines and IBM developed best practices.

Table 6.1 Features included in the Start-Up Bundle

How Start-Up Bundles Work

The following items are the process for the Start-Up Bundle:

- IBM provides you with an SSL certificate and the IP address/hostname of an assigned jump server.

- Using your own Remote Desktop Clients (SSL encrypted), you access the jump server via the Internet. *Note: No permanent network integration into the client environment is required.*

- Your selected bundle is placed into a subnet for the duration of your contract.

- Upon completion of the project, data exports are made available.

Again, to be clear, this entry-level class does not include SAP Basis administration, monitoring for SAP, maintenance of SAP profiles and configuration parameters, SAP updates, or SAP system and client management. Also, there is no Basis-level SAP performance tuning or SAP online support services (OSS) management or automated SAP transport management.

97

CHAPTER 6

As you would also expect, there are no database administration services, nor management of the SAP database(s), periodic reorganization, database-related issue resolution, or update management on the database.

Because this is intended for a non-production environment, there is no batch/interface monitoring or monitoring services for failed jobs and interface connections. With regard to analytics, there are no administration services supporting SAP BusinessObjects applications and components, nor any accompanying services for patching, instance management, output management, and performance tuning in support of current production functionality. There also is no SAP service reporting, so one cannot expect to receive reports for service achievement against service-level agreements (SLAs) or SAP system health review and performance/reliability improvements.

All the above services, however, are available for the Development and Full Service offerings. Should you migrate your Start-Up Bundle system to one of the other two service classes, you receive the above services as well.

Provision-Shared Infrastructure

For the most part throughout this book we've attempted to be generic in describing the process of migrating and SAP system to the cloud. For this section, the overarching concepts are somewhat true of all deployments, but the mechanisms and tools used in what follows are specifically IBM's. Every cloud provider or partner offers similar features, or to be precise, features that accomplish the same ends, but the specifics are different.

Figure 6.5 graphically explains the components and related services behind your cloud implementation.

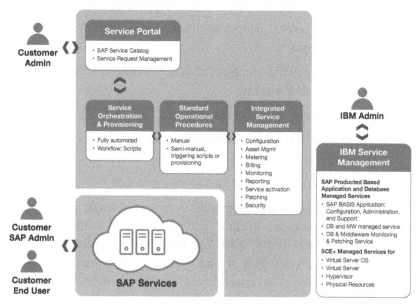

Figure 6.5 Components and organization of the services architecture in the cloud-managed services for SAP applications solution

What follows is an explanation of how IBM sets up your cloud implementation and how you access that system through go-live and beyond. These services include the following:

- **Service Request Portal.** This provides operational management visibility and self-service capabilities for your administrators to view status and request services for all hosted SAP systems and landscapes.

- **Service Management.** This service provides visibility into managed SAP systems and landscapes, monitoring and managing those systems, compliance to SLAs, and billing.

Service Request Portal

The portal provides functions to request the following:

- Creation of new and modification to existing SAP landscapes in the customer cloud environment

- Provisioning of new SAP instances in customer SAP landscapes

- Addition of service extensions to existing SAP systems in customer SAP landscapes

- Modification of the resource entitlements (SAPS, memory, storage) for existing SAP instances

- Performance of one-time service tasks

- Submission of trouble tickets and service requests for existing SAP service instances

- Viewing the status of previous service requests

The portal screens are integrated as part of a single portal system for managing services for all cloud-managed services infrastructure and platform-as-a-service (PaaS) offerings. The portal provides a single user authentication and privileging system with a common look and feel. The Service Request Portal is a web-based central command and control system that provides visibility and control over your SAP environment.

As illustrated in **Figure 6.6**, the self-service portal offers an information dashboard, online service request entry and tracking, online SLA metrics, and bill presentment. Not only does it present a comprehensive and concise view of your cloud implementation, it also serves as the primary channel to enable communication in real-time with your cloud partner when necessary.

Service Management

Remote provisioning, management, and monitoring of SAP service instances in the cloud-managed services is provided by a multi-tier architecture using the IBM Tivoli service management tools installed both locally on each IBM Cloud Managed Services point-of-service delivery data center (POD) and in central management systems as depicted in Figure 6.6.

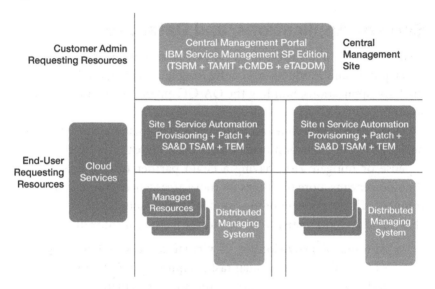

Figure 6.6 Architecture for service management

The central management system is based on the IBM Service Management (ISM) tool, which combines a service catalog of base system images, request fulfillment, incident/problem/change management software, and workflow automation. The shared centralized management system serves as a system of record (SoR), certified for processing financial information and storing all cloud-managed services operations support data such as customer configurations, customer entitlement, incident, problem, and change records. The central management system aggregates data for your asset allocations, metering, and rating information used to calculate billing. Information from multiple local management systems is distributed into the different points of delivery, performing local service management functions.

The distributed managing system has two layers: site (location) level and point-of-delivery-level service. Automation, patching, and service activation and deactivation are managing systems common to an entire site (location). The points-of-delivery service contains a provisioning engine that drives the virtualization managers. Each point of delivery is a self-contained unit of deployment with management software and hardware.

Servers, Applications, and Databases

A sizable chunk of resources in any IT project is building infrastructure—development and production—configuring and loading the OS, building the DEV applications, building the QA/QC environment, and so on, and so on.

That is no longer your responsibility and as this rarely is a core competency for a organization from which it can increase revenue or reduce expense or mitigate risk, having a cloud partner to take this off your shoulders means more resources for those areas where you can build your company. The real benefit of the cloud is that it is standardized, it is virtualized, and it is automated. To reiterate, the features include the following:

- **Automated provisioning of virtual servers and SAP systems**, which reduces the manual tasks required for SAP administration and management to increase the speed of environment creation and reduce the number of errors.

- **Automated processes for service automation, through service activation and deactivation (SA&D),** which at IBM is accomplished using the Tivoli Service Automation Manager (TSAM) and Tivoli Endpoint Manager (TEM). These tools reduce errors and cost associated with activation and deactivation of virtualized resources.

- **Standard operating procedures and project plans for migration** from in-house or hosted SAP installations to the cloud-managed services environment helps to provide SAP systems that are implemented and operated in a consistent manner reducing the downtime, risk, and complexity associated with global SAP implementations.

- **Streamlined workflow, monitoring, and management** provide a high degree of coordination and efficiency, leveraging Tivoli Service Request Manager (TSRM).

- **Globally integrated, standard, repeatable, ITIL-based** processes driven by a workflow service provisioning engine improve speed and flexibility and to provide delivery consistency and quality.

- **Hardened OS images, based on industry practices and processes** adopted from working with partners such as Red Hat and Microsoft.

- **Security based on iSEC policies** and built on secure building blocks result from our experience in outsourcing and hosting.

- **Securely configured middleware based on IBM best practices** and jointly developed security policies.

This is not to say that you do not have assigned responsibilities related to the building phase. Creating the test scripts and providing the testers is probably one of the most important tasks that you should begin immediately. Testing your new cloud environment is where the benefits become more real and concrete, particularly if you use power users who can see those benefits by actually manipulating the system in real time.

Phase Cutover

Once the network is in place and your cloud environment has been built, our best practices recommendation is to perform three migration passes as a standard migration process. The first migration pass, proof of concept, isolates any errors or problems; the second pass acts as a pre-production test; with the final migration pass to production.

CHAPTER 6

CHECKPOINT 6

In this phase, go-live is followed by support and maintenance, and as the following chapter demonstrates, there are areas regarding support and ongoing maintenance that you need to consider. You should have processes and procedures in place prior to go-live.

What we want to talk about before you get into the next chapter is security. To date there have not been any security breaches in any IBM SAP cloud implementation, nor do we expect any because of the extensive security, not only in an SAP system but also in IBM. Not all cloud providers are the same, of course, so in this chapter we explain some of the areas you should discuss with your partner around security.

Security concerns in earlier years were the biggest hurdle to adoption. It became a question of trust, not because the cloud was not secure, but because customers did not have the confidence that they do today. For many, it also was the issue of control. SAP customers who investigated the cloud needed reassurance that they still controlled their data, their applications, and their adoption roadmap (such as when and how to upgrade). Control, however, should never be an issue. After all, it's your system.

What users have access, what data is stored, where in the world the applications and data are hosted—these are big concerns for global corporations. At least as far as IBM is concerned, these are all hardwired into our relationship with our customers. The customer is always right and the customer always decides.

Security is baked into the platform; it is not an add-on. Firewalls around the system and firewalls within the system—between development and production, for example, or between Internet customers and internal users—are available based on the requirements of the customer.

A couple of words about risk. Risk of hacking or similar attacks is much higher if you build your own cloud infrastructure than if you use a partner. If security is a high priority, make it a high priority by becoming informed about the levels of security provided by your cloud partner. In Chapter 7, we'll walk through some of the security features you should confirm with your partner.

The migration to the cloud contains numerous security proto-cols and features. During the Transition phase, security becomes an ongoing practice. This is where the initial extra support takes place and any final documentation for steady state support takes place. The main objective in this phase is to stabilize the system and to close any open items from the production migration. A successful production migration will leave this phase rather unexciting and formal. Nonetheless, it is an important phase as it assures that the new environment knows about the run-time characteristics of the newly migrated workloads and so can provide the correct level of monitoring.

CHAPTER 7

Phase Four: Transition

In this phase, your SAP cloud implementation will transition to its running steady state.

To accomplish this, you and your cloud partner will stabilize the infrastructure, applications, and business processes. Depending on your change management philosophy and best practices, there also will be a training component for your staff and/or partners on the new system.

There also will be a post-migration review with your cloud partner to validate that the new live system is compliant with the functionality as defined by your requirements. Your application management services and infrastructure management services will now be live. Your service-level metrics will be initialized and tracking will begin, along with the methods of reporting the same.

This chapter also will discuss disaster recovery and security, with a focus on how it is managed in a cloud environment by your partner. But before any of this, a few words about go-live.

Go-Live Best Practices

Go-live should be anti-climactic for a cloud implementation. When it comes to SAP HANA, regardless of whether it is on the cloud or on premise, load balancing can be tricky. One of the best things about having a cloud partner is that load balancing becomes its problem, not yours. But make certain you have done your due diligence to make certain your cloud partner has fully taken load balancing into consideration.

If there has been no discussion with your cloud partner of load balancing well before go-live, you may need to make this a priority and make sure they are on top of it. At IBM we seek out load balancing requirements from our clients several weeks before go-live, usually in the form of what we call the load balancer requirements sheet (LBQ).

By the time of go-live, we have identified the demands upon the servers and have fine-tuned your cloud implementation so that there is sufficient capacity to handle the growth and/or unplanned increases to your environment. During the Model phase your cloud partner should already have estimated the demand requirements and during the Build phase should have conducted the necessary testing of the infrastructure, applications, and the success of the data transformation and data conversion.

One of the areas that can affect load balancing is the normal updating and patching of applications and servers, but good news for you, you don't have to do that. That falls to your cloud partner.

One technique for a successful go-live—that is, for production environments—is to conduct a mock cutover. Our usual practice is to conduct a mock cutover and, depending on the number of issues identified, in many cases do it again.

Before any go-live, there needs to be a formal sit-down with your cloud partner, a "go/no-go" meeting. What you don't want to do is to go live without a tete-a-tete with your partner just because your project plan says today's the day. Our best practice is to get a formal sign-off from our customers before go-live.

Service-Level Agreements (SLAs)

SLAs are the foundation for measuring the overall availability and performance of your cloud services. SLAs for cloud-managed services typically include several components, including the following:

- Availability
- System response time
- Service request mean time to respond
- Mean time to address the issue within the service request

While SLAs are a necessity, they are meaningless without an agreed-upon method to track parameters for all committed SLAs and corresponding reports on compliance on a monthly basis via the portal and through ongoing governance meetings. There also needs to be an agreement on what the consequence will be should a service level not be attained—service credits? Discounts? Many cloud providers will request some flexibility on SLAs during the initial stabilization period. You might consider including the stabilization period as a negotiation point in your terms and conditions.

Availability

Availability on the cloud does not mean the servers are up and running. If you can't connect to them or your customers are denied access, the system is not "available." We have heard from some customers that to their cloud partner, "availability" was actually just a euphemism for "up time." Make certain your SLA is specific about describing what availability means. You should require your provider to have application availability SLAs on your production environment that ensure the infrastructure *and* the application are available. This is a much more important SLA and has a bigger impact on your business. Availability also should include the ability of your users to log in, access, and use the SAP environment—application and databases. The virtual machines (VMs) and OS should be operating, and the network infrastructure should support communication to and from the application. In other words, a system is "available" only if users can log in and make productive use of the system.

However, your cloud provider is not responsible for managing your organization's local area network (LAN), so it is important to define the exact point your SLA covers from your organization to the cloud environment. For example, it is not usual to have a network that your cloud partner manages that includes a wide area network (WAN) that connects to your network via routers managed by your cloud partner. In that example, your cloud partner may be responsible for that network but the SLA would not include VPN and other Internet connectivity, as well as your managed network connectivity.

For the purposes of the availability SLA compliance, availability is measured according to the formula shown in **Figure 7.1**.

$$\left(\left(\frac{\text{Minutes of Scheduled Available Time - Minutes of Unscheduled Down Time}}{\text{Minutes of Scheduled Time}}\right)\right) * 100\%$$

Figure 7.1 Formula for availability

Scheduled Available Time: Means 24 hours per day/seven days per week during the applicable contract month.

Unscheduled Downtime: Means any period where the production application environment and/or non-production infrastructure is not available (net of any concurrent downtime among components), excluding scheduled maintenance windows, critical maintenance, customer requested maintenance downtime, and the exclusions.

Your service level for availability should be established based on your requirements. We have three standard service levels for production SAP systems:

- Standard 99.5% availability is included for the production and sandbox SAP systems with Full Service landscapes. For non-production SAP systems, this SLA only applies at the infrastructure level SLA, assuring the availability of the VM, OS, and network virtual LANs (VLANs) in the SAP environment.

- Enhanced 99.7% availability service extension, remaps SAP service VMs to new physical hosts in the cloud infrastructure in the event of cloud host server or hypervisor failure. Reduces allowable application downtime, relative to a standard SLA, by 86 minutes per month.

- High 99.9% availability service extension, allocates redundant VMs located on different physical hosts from the primary SAP service instance VMs. In the event of a server or hypervisor failure of the primary VMs, the SAP service is restarted using the redundant VMs, reducing the allowable application downtime relative to the standard SLA by 173 minutes per month, relative to the standard availability.

Neither the enhanced and high availability configurations are offered for non-production SAP system instances. We think we are pretty good about explaining these standardized options to our SAP customers, but as you can see, each of these options may not appropriate nor as comprehensive as your requirements demand.

Make sure before your go-live that your cloud partner performs an initial availability test along with follow-up testing at regular intervals after go-live.

System Performance/Dialog Response Time

In addition to the business-oriented availability SLAs, your cloud partner also may provide an SLA for system performance in a user-oriented measurement. Production SAP systems user dialog response times are continuously monitored and logged to provide reports demonstrating the quality of the end-user experience. To measure this, you might seek an SLA for the average user dialog response time.

During testing of your new environment, your cloud partner may attempt to define the complexity of your new SAP application components to establish a benchmark to establish the dialog response time SLA. For example, we often establish a system response time SLA of 1,000 milliseconds for SAP online dialog steps, which is measured against SAP standard transaction processing that occurs in the SAP application and database servers. Our clients use the service management tools described in the previous chapter to measure and log all transactions.

Service Delivery SLAs

When issues crop up, your cloud partner should have established SLAs for response times. These SLAs should conform to your requirements. **Table 7.1** shows some sample SLAs to give you a guideline of what you might expect.

CHAPTER 7

Severity class	Description	Response time target value	Resolution time target value
Severity 1	Critical Infrastructure in the case of "Production Application Environment in the case of "Full Service" and/or Application Environment Development Service" has stopped processing. A significant number of users and/or locations are impacted, and the customer's use of the application has stopped or is so severely impacted that customer personnel cannot reasonably continue to work. No workaround, bypass, or alternative is available. A major business impact condition exists. (Critical infrastructure is defined as a network infrastructure, server, or key application outage with a critical impact on service delivery.)	100% <= 15 Minutes Begins upon the go-live Date.	100% <= 5 Hours begins upon completion of the stabilization period
Severity 2	A problem or issue with the customer's "Production" Application (in the case of "Full Service") or with the customer'sApplication (in the case of "Development Service") provided it is operational. However, in the latter case, the following conditions exist: • A substantial feature(s) or key component(s) of the application is down, degraded, or unusable, • Processing is seriously impacted, such that there is a serious impact on the productivity of customer personnel, • Multiple users and/or locations are impacted, • No acceptable workaround, alternative or bypass exists-There is a minimal business impact condition	100% <= 3 hours Begins upon the go-live date.	

Table 7.1 Sample response times for cloud partners

Severity class	Description	Response time target value	Resolution time target value
Severity 3	A problem or issue in which a base component, minor application, or procedure is down, unusable, or difficult to use. There is some or no operational impact and a workaround and/or deferred maintenance is acceptable; or, a less significant aspect (some operational functionality or performance) of an IBM service is unavailable without a workaround. Problems that would be considered Severity Level 1 or 2 that have a workaround, alternative, or bypass available will be assigned a level of Severity 3. Note: All standard Service Requests (RFS).that meet the Severity 2 criteria can be assigned to Severity 2 upon a customer request.	100% <= 2 Business Days Begins upon the commencement of services	

Table 7.1 Sample response times for cloud partners

System Stabilization

Prior to go-live, you should establish with your cloud partner your expectations for the stabilization period of your new SAP system. Remember that one of the major benefits of a cloud implementation is that its time-to-market is normally faster than a traditional SAP implementation. There should be metrics in place related to the inevitable drops in efficiency as a result of user unfamiliarity with the system versus issues with the system itself as your cloud partner makes adjustments to eventually be what is considered a stabilized environment.

As with any SAP system there will be a period during which the production system will not be as productive as the system from which you are migrating. But this period should be relatively short-lived, so it is critical to make certain that you and your cloud partner have not only agreed on the metrics but also on the escalation path should the stabilization period extend beyond the agreed upon expectations.

Operations Process and Training

All organizations have their training standards and protocols. This section is not about training on the application, but on training on how to manage your new cloud environment. As we mentioned in an earlier chapter, your cloud partner will have some interface for you to track your SLAs, application usage, cloud environment, and more. In IBM's case, we provide a portal.

Portal

The IBM Cloud for SAP Applications portal (the "portal") is a secure, Internet-native customer service portal for managing your relationship with your cloud delivery teams. The portal is designed to provide you control over your environments. The portal includes a self-service catalog of services that allows customers to submit service requests directly and view the status (new, queued, in-progress, closed, or fulfilled) of those requests.

"Cloud Delivery" is a managed application services (MAS) self-service portal (the "portal") to manage and deliver availability, reliability, security, and scalability for your hosted environments. The portal supports service request management, change management, and problem and incident management. In IBM's model, we train you on how to use the portal.

Security via Authorized Contacts

Portals support day-to-day communication between the SAP customer and the cloud provider, but to ensure that only properly credentialed staff are making updates, changes, and queries to your cloud implementation, it is critical that proper protocols be in place. IBM's approach is to only accept service requests submitted into the managed services portal application by authorized contacts, or received via telephone from authorized contacts.

You get to decide who your authorized contacts will be. Note that there may be several authorized contacts depending upon a particular situation or role (we recognize a wide range of business manager roles and customer user roles, for example).

Your authorized contacts may include one or several of the following roles:

- **Customer Call Center Representative.** You may have one or more individuals within your call center or help desk who are the first point of contact for any internal IT request.

- **Customer Super User.** As part of their core job responsibilities, customer "super users" utilize the hosted environments every day and are knowledgeable in their functionality as well as the business processes of your company.

- **Service Request Manager or Change Control Manager.** You may have a dedicated service request manager or change control manager who coordinates all system changes for your organization.

- **Systems Integrator Representative.** During projects, you may have a system integrator who needs to interact extensively with your cloud partner. At IBM, we only allow authorized access to system integrators based on assignment from you.

Disaster Recovery

In a previous chapter, we wrote at length about identifying whether the system you are replacing (or better, the business processes you are replacing or extending into the cloud) is mission critical or not.

Well before go-live, you need to establish what level of disaster recovery you require. Most cloud partners provide a disaster recovery remedy based on the recovery point objective/recovery time objective (RPO/RTO) in your requirements. For example, your requirements may state that in the case of a crash or similar disaster, you want your system to have a recovery point objective of 15 minutes and a recovery time objective of four hours.

CHAPTER 7

Regardless of your requirements, it will be up to your cloud provider to determine and present a solution. The technology and the parameters of the disaster recovery solution are the result of the combination of the customer requirements and the solutions proposed by IBM.

Figure 7.2 illustrates a simplified view of protecting your cloud managed service (CMS) with either a mirror or log shipping solution to a secondary data center.

Figure 7.2 Disaster recovery layout

There are a variety of technologies available for disaster recovery of SAP cloud environments. Two used by IBM include Global Mirror (an IBM technology) and log shipping. Global Mirror, like other mirroring technologies, updates a backup copy of your SAP implementation at regular intervals, usually determined by the SLA. Log shipping, a more economical option, creates an automated backup of database and transaction log files from a production database. That backup can be moved into production quickly. Both options include both failover and failback.

As part of the SLA negotiation, you should establish with your cloud partner which disaster recovery method you prefer, based on your requirements. Your cloud partner may only offer a certain solution based on the hardware, OS, and applications that support your cloud implementation. Disaster recovery options that may be up for negotiation include the following:

- Frequency of testing of the system. Annually? Semi-annually?

- Workload testing, including how often. Quarterly is a common interval.

In **Table 7.2**, you'll find an example of a disaster recovery rubric between a cloud partner and a customer.

DR technology	Constraints	RPO/RTO
Global Mirror Data Replication	• Only for applications • Specific sites only	• RPO one (1) hour/RTO four (4) hours for first two (2) SAP Applications • RPO one (1) hour/RTO six (6) hours for SAP Applications three (3) and four (4) • RPO one (1) hour/RTO eight (8) hours for SAP Applications five (5) and six (6) • RPT one (1) hour/RTO ten (10) hours for SAP Application seven (7) and eight (8)
Database redo log shipping	Specific sites only	• RPO one (1) hour/RTO four (4) hours for the first two (2) SAP Applications • RPO one (1) hour/RTO six (6) hours for SAP Applications three (3) and four (4) • RPO one (1) hour/RTO eight (8) hours for SAP Applications five (5) and six (6) • RPT one (1) hour/RTO ten (10) hours for SAP Application seven (7) and eight (8)

Table 7.2 An example of a disaster recovery rubric between a cloud partner and a customer

Security

While SAP has a good reputation for security, it goes without saying that this should still be an area where, when it comes to the cloud, you need to be vigilant. What follows is some general advice regarding security and your cloud partner that you should find will help when crafting a request for information (RFI) or request for proposal (RFP) of potential cloud partners.

Cloud Infrastructure Security

Your SAP cloud provider possesses security features to protect your cloud implementation infrastructure from network threats, securing the environment from which SAP services are delivered. The cloud security model should proactively adopt security strategies and capabilities to harden each layer of the solution by focusing on continuous improvement of security processes. Your cloud partner must maintain a security-rich environment, from both a physical and logical perspective.

IBM's approach has been to develop a unified security framework, over which the technologies and services are delivered. The features of our security foundation integrate key security-rich technologies at each layer of the solution. When mulling your security requirements, here are some features your cloud partner should include the following:

- Perimeter security, including firewall and intrusion detection systems.
- Managed security event monitoring and reporting, combining multiple security devices.
- Vulnerability assessment, penetration testing, and incident response management and reporting.
- Layered anti-virus protection.
- Automated asset discovery and patch management.
- Pervasive auditing of vulnerabilities, virus protection, and operating system patches.
- Physical security and access, in this case of your cloud partner's facilities. At IBM, we use badge access to the facilities granted only to IBM personnel and authorized contractors. Critical infrastructure, including servers, security infrastructure, and networking equipment, is housed in secure facilities protected by steel cages, double-door entry ways, and hallway cameras.

You should also investigate how your cloud partner addresses security awareness, the so-called "human perimeter," which elevates the importance of security-rich processes and employee responsibility through awareness, education, process definition, and certified security experts on staff. How often are employees screened? And if so, do those screenings include the following:

- Verification of previous employment
- Verification of education and technical certifications
- Criminal background checks, both federal and civil
- Verification of right-to-work
- Debarment check, authorization to work on government contracts
- Social Security checks for the prior seven years
- Residences for the prior seven years

Your cloud partner also should perform external security assessments. Today there is a wide variety of security assessment software, tools, and scripts. The assessments should focus on all layers of the technology stack including the operating system, the web application, and the database. Areas to be examined should include the following:

- Network topology, vulnerabilities, and configurations of devices

- Discovery of the business impacts (if any) of possible attack scenarios, exploitable vulnerabilities, and policy violations

- Reporting of ranked business risks and implementation of cost-effective remedies

Management Layer Security

The management layer is used to manage both the infrastructure and SAP environments. The management layer security includes the following:

- Application network isolation using physical and logic separation, secure trunking and channeling, and VLANs

- Management networking separate from SAP Application networking for access to the management infrastructure

- Storage networking separated using both zoning and hypervisor isolation

- Regular validation of security parameters and policies

- Security patching and vulnerability scanning

- International Organization for Standardization (ISO)/International Electrotechnical Commission (IEC) 27001/2-based security policy that supports industry and regulatory requirements

- Hardened OS images

- Securely-configured middleware, based on security policy specifications

- Automated validation against ISeC security controls

CHAPTER 7

Security Measures

SAP environments should be built in a logically partitioned virtual data center provisioned on the CMS infrastructure. SAP applications run on VMs, each with its own distinct, dedicated database. Security services provided should include design and operation elements in the following areas:

- SAP Identity and Access Management
- SAP Patch Management
- SAP Log Management
- SAP Health Checking
- SAP Service Activation and Deactivation Procedures

SAP systems are secured according to the security policies and procedures defined for their own internal critical applications and infrastructure. Your cloud provider should maintain security policy management for the SAP application environment for all hosted systems based on best practices. We have our own proprietary IBM Information Security Controls (ISeC) standards that define standard security policies for privileged authorizations, user ID management, password requirements, logging requirements, and other technical controls. These policies and controls are implemented both in the SAP application level and in the underlying infrastructure on which the SAP application components are hosted.

The ISeC standard defines minimum security parameters that will be automatically enforced by software controls integrated into the cloud hosting environment. You can require different parameter settings for the given set of SAP technical controls as long as they are above the established and defined minimum level.

Whatever security compliance support services your cloud provider uses, you need to make certain they perform periodic internal security reviews to audit and validate compliance to those standards.

Again, using IBM as an example, we perform system security checking that verifies the following criteria:

- Anti-virus software is functional and operating on VMs.
- Software tools for security policy management are functional and operate as defined in the documented ISeC standards.

- IBM will perform security-related system currency.

- Based on the automated OS patching process services defined in the Cloud Managed Services for SAP Applications base service.

- Based on manual installed SAP Basis and database management system (DBMS) security-relevant patches and/or support packages.

IBM will perform the management of privileged user IDs and IBM user IDs for OS, DBMS, and SAP Basis components.

Don't Forget Industry-Specific Security

Corporations in the healthcare and pharmaceutical industries need to be concerned with the Health Insurance Portability and Accountability Act of 1996 (HIPAA) privacy rules. Corporations that rely on credit card transactions need to maintain Payment Card Industry (PCI) compliance. Data, be it patient medical histories or debit card numbers, requires specialized or enhanced security.

CHAPTER 7

CHAPTER 7

Notes

Notes